Thomas W. Wieting
Reed College
1984

HAUSDORFF MEASURES

HAUSDORFF MEASURES

C. A. ROGERS, D.Sc., F.R.S.

Astor Professor of Mathematics
University College London

CAMBRIDGE
AT THE UNIVERSITY PRESS
1970

Published by the Syndics of the Cambridge University Press
Bentley House, 200 Euston Road, London N.W.1
American Branch: 32 East 57th Street, New York, N.Y.10022

© Cambridge University Press 1970

Library of Congress Catalogue Card Number: 74–123674

Standard Book Number: 0 521 07970 5

Printed in Great Britain
at the University Printing House, Cambridge
(Brooke Crutchley, University Printer)

CONTENTS

PREFACE

Measures are of importance in mathematics in two rather different ways. Measures can be used for estimating the size of sets, and measures can be used to define integrals. E. Borel in his 1894 thesis (see Borel 1895 or 1940)† essentially introduced the Lebesgue outer measure as a means of estimating the size of certain sets, so that he could construct certain pathological functions. Lebesgue (1904) on the other hand was mainly interested in his measures as a tool enabling him to construct his integral. While both aspects of measure theory are important, the emphasis in this book will be almost entirely on the first; we only mention the theory of integration in the last section of the last chapter.

The first 'Hausdorff' measure was introduced by C. Carathéodory (1914), in a paper in which he also introduced the much more general Carathéodory outer measures. Carathéodory developed the theory of linear measure in n-dimensional Euclidean space and in a final paragraph clearly showed how p-dimensional measure could be introduced in q-dimensional space, for $p = 1, 2, \ldots, q$. The p-dimensional measures, for general positive real p, were introduced by F. Hausdorff (1919); he also illustrated the use of these measures by showing that the Cantor ternary set has in a certain sense the fractional dimension $\log 2/\log 3 = 0{\cdot}6309 \ldots$. The theory of Hausdorff measures has developed very greatly since 1919, very largely as a result of the work of A. S. Besicovitch and his students.

This book cannot contain an account of more than a tiny fraction of the work that has been done on Hausdorff measures. After a first chapter giving an introduction to measure theory with special attention to the study of non-σ-finite measures, the second chapter develops the most general aspects of the theory of Hausdorff measures, and the final chapter gives an account of the applications of Hausdorff measures, a general survey being followed by detailed accounts of two rather special topics.

Much of this book is based on postgraduate lectures given at the University of British Columbia and at University College London. I am conscious that I have been considerably influenced by close and enjoyable contacts with Maurice Sion, when I visited the University

† Such references are to the bibliography, p. 169.

of British Columbia, and with Roy O. Davies, when he visited University College London. Indeed I must hasten to acknowledge that much of §7 of Chapter 2 is the result of joint work with Dr Davies, and that I am most grateful to him for permission to publish it first in this work. Dr Davies has also helped me by careful criticism of Chapter 1. As the book is largely based on lectures, and, as I like my students to follow my lectures, proofs are given in great detail; this may bore the mature mathematician, but it will I believe be a great help to anyone trying to learn the subject *ab initio*.

C. A. ROGERS

November, 1969

I am most grateful to the Cambridge University Press for the care they have taken in the production of this book, and to Miss S. Burrough for the assistance she has given with the proof-reading.

C. A. R.

July, 1970

1

MEASURES IN ABSTRACT, TOPOLOGICAL AND METRIC SPACES

§1 Introduction

In his fundamental book H. Lebesgue (1904) introduces an exterior measure or outer measure $m_e(E)$ and an interior measure or inner measure $m_i(E)$ associated with *every* subset E of the real line. He then associates a measure $m(E)$, the common value of $m_e(E)$ and $m_i(E)$, with those sets for which $m_e(E) = m_i(E)$. When more general measures were studied by J. Radon (1913) and C. Carathéodory (1914), these authors adopted slightly different points of view. Radon placed his emphasis on the measures of countably additive set functions defined on a Borel ring of sets and Carathéodory concentrated his attention on the outer measure defined for all sets. In this chapter we adopt the Carathéodory point of view, in this, as in many other ways; but we take care to explain the relationship between the two points of view and to show that they are essentially equivalent.

We make no attempt to present a complete account of any aspect of measure theory, giving only an introduction developing those parts of the general theory that are useful for the study and appreciation of Hausdorff measures. In view of this purpose we take care to prove, in as far as this is possible, results that apply to non-σ-finite measures.

§2 Measures in abstract spaces

In this section we develop the theory of measures in abstract spaces from the 'outer measure' point of view. We first define measures and give a number of examples. We then introduce the concept of measurability and develop the elementary properties of measurable sets. We then introduce 'Method I', a method of defining a measure from a set function satisfying only the weakest of conditions, and show that every measure can be regarded as a measure constructed in this way. We then establish the relationships between the 'outer measure' approach to measure theory and the equally valid approach through measures defined on σ-fields of sets. This leads us to a study of regular measures. Finally, we give methods of obtaining a measure by relativizing a given measure and by taking the supremum of a family of measures.

As soon as one agrees to call the elements of a set 'the points' of the set, the set becomes an abstract space. In this section we shall work with such a space Ω. We introduce

***Definition* 1.** *A function μ defined on the sets of a space Ω is called a measure on Ω if it satisfies the conditions:*

(*a*) *$\mu(E)$ is a non-negative real number or $+\infty$ for each sub-set E of Ω;*

(*b*) *$\mu(\varnothing) = 0$;*

(*c*) *if $E_1 \subset E_2$ then $\mu(E_1) \leqslant \mu(E_2)$; and*

(*d*) *if $\{E_i\}$ is any sequence of sets of Ω then*

$$\mu\left(\bigcup_{i=1}^{\infty} E_i\right) \leqslant \sum_{i=1}^{\infty} \mu(E_i).$$

Here we depart from standard practice; most authors call the set functions satisfying these conditions 'outer measures' and reserve the name 'measure' for the restriction of such an 'outer measure' to a σ-field (see definition 3 below) of sets on which it is countably additive (see definition 4 below).

Even in a completely abstract space it is easy to give examples of such measures.

Example A. If E is a finite set of points take $\mu(E)$ to be the number of points of E, if E is an infinite set of points take $\mu(E) = +\infty$. We call this measure 'counting measure'.

Example B. Choose a fixed infinite cardinal and take $\mu(E) = 0$, if the cardinal of E does not exceed the fixed cardinal, and take

$$\mu(E) = +\infty,$$

if the cardinal of E does exceed the fixed cardinal.

The second example illustrates a general method of passing from a suitable set property to a measure. Suppose we have a set property S that satisfies the conditions:

(*a*) $\varnothing$ has S;

(*b*) if E_1 has S and $E_2 \subset E_1$ then E_2 has S;

(*c*) if $\{E_i\}$ is a sequence of sets having S then $\bigcup_{i=1}^{\infty} E_i$ has S.

Here the property S can be read as meaning that E is small, in some sense, if E has S. We get examples of measures by taking such a property.

Example C. Let S be a set property satisfying (a), (b) and (c). Take $\mu(E) = 0$, if E has S, and $\mu(E) = 1$, if E has not got S.

Example D. Let S be a set property satisfying (a), (b) and (c). Take $\mu(E) = 0$, if E has S, and $\mu(E) = +\infty$, if E has not got S.

It is easy to give examples of set properties satisfying (a), (b) and (c).

(i) E is at most countable.

(ii) E has cardinal not exceeding some fixed infinite cardinal.

(iii) When Ω is a topological space; E is a countable union of nowhere dense sets (i.e. E is of the first category).

(iv) When Ω is a Hausdorff space; E is a subset of some countable union of compact sets.

(v) When Ω is a topological space and n is given; E is a countable union of sets of topological dimension at most n, for any reasonable definition of topological dimension.

(vi) When a measure ν is given, E has $\nu(E) = 0$.

(vii) When a measure ν is given, E is a countable union of sets of finite ν measure.

Coming down to n-dimensional Euclidean space R_n we have the well-known examples.

Example E. When Ω is R_n; let $\mu(E)$ be the Lebesgue (outer) measure of E (see §5 below).

Example F. When Ω is R_n; let $\mu(E)$ be the upper Lebesgue–Stieltjes or Radon integral

$$\overline{\int_E} dF(x),$$

where F is a non-negative interval function (see Radon, 1913).

2.1. While our measures are defined on all the sets of the space they have rather special properties on the class of 'measurable' sets. We work with a variant of Carathéodory's definition.

Definition 2. *If μ is a measure on Ω, a set E is said to be μ-measurable if for all sets A, B with*

$$A \subset E, \quad B \subset \Omega \backslash E \tag{1}$$

we have

$$\mu(A \cup B) = \mu(A) + \mu(B). \tag{2}$$

Here we use the symbol '$\backslash$' between two sets to mean 'set theoretic difference'; $A \backslash B$ meaning the set of those points of A not in B (this can be read 'A less B').

We say that sets A, B satisfying (1) are separated by E; the definition takes the form: *E is μ-measurable if μ is additive on sets that are separated by E.*

As we shall often have to check the measurability of sets it is convenient to have a measurability criterion that is slightly easier to use.

Theorem 1. *If μ is a measure on Ω, then a set E is μ-measurable, if we have*

$$\mu(A \cup B) \geqslant \mu(A) + \mu(B), \tag{3}$$

whenever A and B are sets of finite μ-measure that are separated by E.

Proof. Applying the defining property (d) of a measure to the sequence $A, B, \varnothing, \varnothing, \ldots$ and using the property (b) we have

$$\mu(A \cup B) \leqslant \mu(A) + \mu(B)$$

for all sets A, B. So, to prove the equality (2) for all sets A, B separated by E, it suffices to prove that the inequality (3) holds whenever one of the sets A, B separated by E has infinite measure. But in this case the defining property (c) ensures that both sides of (2) have the value $+\infty$. Hence E is μ-measurable.

Corollary. *If the only values taken by the measure μ are 0 or ∞, all sets are μ-measurable.*

Proof. Whenever A and B have finite μ-measure

$$\mu(A \cup B) \geqslant 0 = \mu(A) + \mu(B).$$

2.2. We are now in a position to state a theorem giving considerable information about the measurable sets and the behaviour of the measure on the measurable sets.

Theorem 2. *Let μ be a measure on Ω. Then*

(a) *if $\mu(N) = 0$, N is μ-measurable;*

(b) *if E is μ-measurable, so is $\Omega \backslash E$;*

(c) *if $\{E_i\}$ is a sequence of μ-measurable sets, $\bigcup_{i=1}^{i=\infty} E_i$ and $\bigcap_{i=1}^{i=\infty} E_i$ are μ-measurable;*

(d) *if $\{E_i\}$ is a disjoint sequence of μ-measurable sets,*

$$\mu\left(\bigcup_{i=1}^{\infty} E_i\right) = \sum_{i=1}^{\infty} \mu(E_i).$$

Before we prove this theorem it will be convenient to introduce the concept of a σ-field† and to prove a simple lemma on σ-fields.

† Read 'sigma-field'.

***Definition* 3.** *A system $\mathscr{A}$ of sets is called a σ-field if it has the three properties:*

(*a*) $\varnothing \in \mathscr{A}$;
(*b*) *if* $A \in \mathscr{A}$, *then* $\Omega \setminus A \in \mathscr{A}$;
(*c*) *if* $A_i \in \mathscr{A}$ *for* $i = 1, 2, \ldots,$ *then* $\bigcup_{i=1}^{i=\infty} A_i \in \mathscr{A}$.

***Lemma* 1.** *Let $\mathscr{A}$ be a system of sets with the four properties:*

(*a*) $\varnothing \in \mathscr{A}$;
(*b*) *if* $A \in \mathscr{A}$, *then* $\Omega \setminus A \in \mathscr{A}$;
(c_1) *if* $A_1 \in \mathscr{A}$ *and* $A_2 \in \mathscr{A}$ *then* $A_1 \cup A_2 \in \mathscr{A}$;
(c_2) *if* $A_i \in \mathscr{A}$ *for* $i = 1, 2, \ldots,$ *and* $A_1, A_2, \ldots,$ *are disjoint, then* $\bigcup_{i=1}^{i=\infty} A_i \in \mathscr{A}$.

Then $\mathscr{A}$ is a σ-field.

Proof. By conditions (*a*) and (*b*) the corresponding conditions of definition 3 hold. So we have only to establish condition (*c*) of the definition. Let $A_1, A_2, \ldots$ be any sequence of sets of $\mathscr{A}$. Then we can write

$$\bigcup_{i=1}^{\infty} A_i = \bigcup_{i=1}^{\infty} [A_i \cap \{\Omega \setminus \bigcup_{j<i} A_j\}]. \tag{4}$$

By condition (*a*) when $i = 1$, and trivially when $i = 2$ and by condition (c_1), iterated when necessary, when $i \geqslant 3$, the sets

$$\bigcup_{j<i} A_j \quad (i = 1, 2, \ldots)$$

all belong to $\mathscr{A}$. Using (*b*) and (c_1) it follows that the sets

$$A_i \cap \{\Omega \setminus \bigcup_{j<i} A_j\} = \Omega \setminus [\{\Omega \setminus A_i\} \cup \{\bigcup_{j<i} A_j\}]$$

belong to $\mathscr{A}$. As these sets are disjoint sets of $\mathscr{A}$, it follows from (c_2) and (4) that $\bigcup_{i=1}^{i=\infty} A_i$ belongs to $\mathscr{A}$, as required.

Remark. The formula

$$\bigcap_{i=1}^{\infty} A_i = \Omega \setminus \left[\bigcup_{i=1}^{\infty} \{\Omega \setminus A_i\}\right]$$

shows that a σ-field is automatically closed under the operation of countable intersection.

Proof of theorem 2. The proof will be given in parts.

Proof of (*a*). Suppose that N is a set with $\mu(N) = 0$ and that A, B are separated by N with $A \subset N$, $B \subset \Omega \setminus N$. Then using the defining properties in the order (*c*), (*d*), (*c*) and (*b*), we obtain

$$\begin{aligned}\mu(B) \leqslant \mu(A \cup B) &\leqslant \mu(A)+\mu(B)\\ &\leqslant \mu(N)+\mu(B)\\ &= \mu(B).\end{aligned}$$

Hence we must have equality throughout and N is μ-measurable.

Proof of (*b*). Let E be μ-measurable. Then $\Omega\backslash E$ is also μ-measurable by the symmetry of the definition in E and $\Omega\backslash E$.

Proof of (c) *for the union of two sets.* Let E_1, E_2 be two μ-measurable sets. Let A and B be any two sets with finite μ-measure and with

$$A \subset E_1 \cup E_2, \quad B \subset \Omega\backslash(E_1 \cup E_2).$$

Now
$$A \cup B = [A \cap E_1] \cup [\{A \cup B\} \cap \{\Omega\backslash E_1\}]$$

and the sets $A \cap E_1$ and $\{A \cup B\} \cap \{\Omega\backslash E_1\}$ are separated by the measurable set E_1. Hence

$$\mu(A \cup B) = \mu(A \cap E_1) + \mu(\{A \cup B\} \cap \{\Omega\backslash E_1\}). \tag{5}$$

But
$$\{A \cup B\} \cap \{\Omega\backslash E_1\} = \{A \cap (\Omega\backslash E_1)\} \cup B$$

and the sets $A \cap (\Omega\backslash E_1)$ and B are separated by the measurable set E_2. Hence

$$\mu(\{A \cup B\} \cap \{\Omega\backslash E_1\}) = \mu(A \cap (\Omega\backslash E_1)) + \mu(B). \tag{6}$$

Finally, as E_1 is measurable

$$\mu(A \cup E_1) + \mu(A \cap (\Omega\backslash E_1)) = \mu(A). \tag{7}$$

So, by (5), (6) and (7) we have

$$\mu(A \cup B) = \mu(A) + \mu(B)$$

and $E_1 \cup E_2$ is μ-measurable.

Proof of (c) *for the union of a disjoint sequence of μ-measurable sets and proof of* (d). *Let* $\{E_i\}$ be a disjoint sequence of μ-measurable sets. Write

$$E = \bigcup_{i=1}^{\infty} E_i$$

and let A and B be sets with

$$A \subset E \quad \text{and} \quad B \subset \Omega\backslash E.$$

By repeated application of the result of the last paragraph, the set $\bigcup_{i=1}^{i=n} E_i$ is μ-measurable for each positive integer n. Hence

$$\begin{aligned}\mu(A \cup B) &\geqslant \mu\left(\left[A \cap \left\{\bigcup_{i=1}^{n} E_i\right\}\right] \cup B\right)\\ &= \mu\left(A \cap \left\{\bigcup_{i=1}^{n} E_i\right\}\right) + \mu(B),\end{aligned} \tag{8}$$

as
$$B \subset \Omega \backslash E \subset \Omega \backslash \left\{ \bigcup_{i=1}^{n} E_i \right\}.$$

Since the sets $E_n, E_{n-1}, \ldots, E_1$ are disjoint and μ-measurable, we have

$$\begin{aligned}
\mu\left(A \cap \left\{ \bigcup_{i=1}^{n} E_i \right\}\right) &= \mu\left(\left[A \cap \left\{ \bigcup_{i=1}^{n-1} E_i \right\}\right] \cup [A \cap E_n]\right) \\
&= \mu\left(A \cap \left\{ \bigcup_{i=1}^{n-1} E_i \right\}\right) + \mu(A \cap E_n) \\
&= \mu\left(A \cap \left\{ \bigcup_{i=1}^{n-2} E_i \right\}\right) + \mu(A \cap E_{n-1}) + \mu(A \cap E_n) \\
&= \ldots \\
&= \mu(A \cap E_1) + \mu(A \cap E_2) + \ldots + \mu(A \cap E_n).
\end{aligned}$$

Using this in (8), we obtain
$$\mu(A \cup B) \geqslant \sum_{i=1}^{n} \mu(A \cap E_i) + \mu(B).$$

As this holds for each integer n, it implies that
$$\mu(A \cup B) \geqslant \sum_{i=1}^{\infty} \mu(A \cup E_i) + \mu(B).$$

So, using the defining property (d) of a measure,
$$\begin{aligned}
\mu(A \cup B) &\geqslant \sum_{i=1}^{\infty} \mu(A \cap E_i) + \mu(B) \\
&\geqslant \mu\left(A \cap \left\{ \bigcup_{i=1}^{\infty} E_i \right\}\right) + \mu(B) \\
&= \mu(A) + \mu(B) \\
&\geqslant \mu(A \cup B). \qquad (9)
\end{aligned}$$

In the first place, this shows that $\mu(A \cup B) \geqslant \mu(A) + \mu(B)$ so that $\bigcup_{i=1}^{i=\infty} E_i$ is μ-measurable. Secondly, taking the special case when $B = \varnothing$, and noting that equality must hold throughout (9), we obtain
$$\mu(A) = \sum_{i=1}^{\infty} \mu(A \cap E_i), \qquad (10)$$

for all subsets A of $E = \bigcup_{i=1}^{i=\infty} E_i$. Thirdly, taking $A = E$ in (10) we obtain
$$\mu\left(\bigcup_{i=1}^{\infty} E_i\right) = \sum_{i=1}^{\infty} \mu(E_i). \qquad (11)$$

Thus we have established (d) and also (c) for the union of a disjoint sequence of μ-measurable sets.

Proof of (c). By (a), the set $\varnothing$ is μ-measurable. Further we have already shown that the system of μ-measurable sets is closed under the operations of complementation with respect to Ω, of union of two sets, and a countable disjoint union. It follows, by Lemma 1 that this system is a σ-field and so is closed under the operations of countable union and intersection. This completes the proof of the theorem.

Corollary. *Let* T *be any set in* Ω. *Let* $\{E_i\}$ *be a disjoint sequence of* μ*-measurable sets. Then*

$$\mu\left(T \cap \bigcup_{i=1}^{\infty} E_i\right) = \sum_{i=1}^{\infty} \mu(T \cap E_i). \tag{12}$$

Proof. Applying (10) with

$$A = T \cap \bigcup_{i=1}^{\infty} E_i,$$

we obtain

$$\mu\left(T \cap \bigcup_{i=1}^{\infty} E_i\right) = \sum_{i=1}^{\infty} \mu\left(T \cap \left\{\bigcup_{j=1}^{\infty} E_j\right\} \cap E_i\right) = \sum_{i=1}^{\infty} \mu(T \cap E_i),$$

as required.

2.3. At this stage it is convenient to introduce the concept of a countably additive measure defined on a σ-field of sets.

***Definition* 4.** *A set function* μ *defined on a* σ*-field* $\mathscr{A}$ *of sets is called a countably additive measure on* $\mathscr{A}$ *if it has the following properties*:

(a) $0 \leqslant \mu(A) \leqslant +\infty$ *for all* A *in* $\mathscr{A}$;

(b) $\mu(\varnothing) = 0$;

(c) *whenever* $\{A_i\}$ *is a disjoint sequence of sets of* $\mathscr{A}$,

$$\mu\left(\bigcup_{i=1}^{\infty} A_i\right) = \sum_{i=1}^{\infty} \mu(A_i).$$

Remark. Many authors reserve the name 'measure' for such countably additive measures defined on a σ-field of sets.

In terms of this definition we can reword Theorem 2 to yield

Theorem 3. *If* μ *is a measure on* Ω, *the system* $\mathscr{M}$ *of* μ*-measurable sets is a* σ*-field containing the null sets (i.e. those sets* N *with* $\mu(N) = 0$), *and the restriction of* μ *to* $\mathscr{M}$ *is a countably additive measure on* $\mathscr{M}$.

2.4. The relationship between the measure μ defined on all the subsets of Ω and its restriction to its measurable sets is close. Given a countably additive measure ν defined on a σ-field $\mathscr{A}$ we can, by a process that we shall shortly describe, extend ν to yield a measure λ defined on all sets of Ω, with

$$\lambda(A) = \nu(A) \quad \text{for} \quad A \in \mathscr{A},$$

in such a way that $\mathscr{A}$ is a subfield of the σ-field of λ-measurable sets. In the special case, when ν is the restriction of a measure μ defined on Ω to the μ-measurable sets, and μ is a regular measure (see Theorem 7 below), the measure λ generated from ν will coincide with μ.

For this particular purpose we need a method of constructing a measure from a countably additive measure defined on a σ-field. For other reasons it is desirable to have a much more general method of constructing a measure from a 'pre-measure' defined on any class of sets that contain $\varnothing$. Following Munroe (1953) we shall call this method 'Method I'.

***Definition* 5.** *A function τ defined on a class $\mathscr{C}$ of subsets of Ω will be called a pre-measure, if:*

(*a*) $\varnothing \in \mathscr{C}$;

(*b*) $0 \leqslant \tau(C) \leqslant +\infty$ *for all C in $\mathscr{C}$*;

(*c*) $\tau(\varnothing) = 0$.

***Theorem* 4.** *If τ is a pre-measure defined on a class $\mathscr{C}$ of sets, the set function*

$$\mu(E) = \inf_{\substack{C_i \in \mathscr{C} \\ \cup C_i \supset E}} \sum_{i=1}^{\infty} \tau(C_i) \tag{13}$$

is a measure on Ω.

Remarks. We adopt the convention that any infimum taken over an empty set of real numbers has the value $+\infty$. When no confusion can arise we will abbreviate the formula (13) to the form

$$\mu(E) = \inf_{\cup C \supset E} \Sigma\, \tau(C_i). \tag{14}$$

We shall call the measure μ the measure constructed from the pre-measure τ by Method I. We can call a sequence $\{\mathscr{C}_i\}$ of sets of $\mathscr{C}$ with $\bigcup_{i=1}^{i=\infty} C_i \supset E$ a covering of E with sets from $\mathscr{C}$ and we can call the sum $\sum_{i=1}^{i=\infty} \tau(C_i)$ the τ-value of the covering. Then $\mu(E)$ is the infimum of the τ-values of the coverings of E by sets from $\mathscr{C}$. Our notation conventionally excludes the possibility of using a finite system of sets

$$C_1, C_2, \ldots, C_n$$

of $\mathscr{C}$ covering E; but given such a covering we can replace it by the covering $\{D_i\}$ with

$$D_i = C_i \quad (i = 1, 2, \ldots, n),$$

$$D_i = \varnothing \quad (i = n+1, n+2, \ldots),$$

and we then have

$$\sum_{i=1}^{\infty} \tau(D_i) = \sum_{i=1}^{n} \tau(C_i).$$

Proof. (*a*) As $0 \leqslant \tau(C) \leqslant +\infty$ for all C in $\mathscr{C}$, we clearly have

$$0 \leqslant \mu(E) \leqslant +\infty$$

for all E in Ω.

(*b*) We have $\mu(\varnothing) = \inf_{\cup C \supset \varnothing} \Sigma\tau(\mathscr{C}_i) \leqslant \Sigma\tau(\varnothing) = 0.$

Hence $\mu(\varnothing) = 0$.

(*c*) If $E_1 \subset E_2$, any cover of E_2 also covers E_1 and so

$$\mu(E_1) \leqslant \mu(E_2).$$

(*d*) Let $\{E_i\}$ be any sequence of sets of Ω. We prove that

$$\mu\left(\bigcup_{i=1}^{\infty} E_i\right) \leqslant \sum_{i=1}^{\infty} \mu(E_i). \tag{15}$$

This result is trivial if $\sum_{i=1}^{\infty} \mu(E_i) = +\infty.$

So we may suppose that $\sum_{i=1}^{\infty} \mu(E_i)$

is finite. Then, in particular, each $\mu(E_i)$ is finite. So, if $\epsilon > 0$ is given, for each integer $i > 1$ we can choose a sequence $\{C_j^{(i)}\}_{j=1}^{j=\infty}$ of sets in $\mathscr{C}$ with

$$E_i \subset \bigcup_{j=1}^{\infty} C_j^{(i)},$$

$$\sum_{j=1}^{\infty} \tau(C_j^{(i)}) \leqslant \mu(E_i) + \epsilon . 2^{-i}.$$

Let $\{D_i\}$ be a sequence obtained by rearranging the sets $C_j^{(i)}$ $(i, j = 1, 2, \ldots)$ as a single sequence. Then

$$\bigcup_{i=1}^{\infty} E_i \subset \bigcup_{i=1}^{\infty} D_i,$$

$$D_i \in \mathscr{C} \quad (i = 1, 2, \ldots)$$

and
$$\mu\left(\bigcup_{i=1}^{\infty} E_i\right) \leqslant \sum_{i=1}^{\infty} \tau(D_i)$$
$$= \sum_{i,j=1}^{\infty} \tau(C_j^{(i)})$$
$$\leqslant \sum_{i=1}^{\infty} \{\mu(E_i) + \epsilon . 2^{-i}\}$$
$$= \left\{\sum_{i=1}^{\infty} \mu(E_i)\right\} + \epsilon.$$

As ϵ may be any positive number, this proves (15).

We have now checked the four defining properties for a measure.

2.5. In our next theorem we show that any measure μ defined on Ω can be obtained by applying Method I to a suitable pre-measure. While this theorem establishes the generality of the method it distracts attention from the main use of the method which is to construct a measure μ defined on all the sets of Ω from a pre-measure τ defined, preferably in a simple way, on a relatively small class of subsets of Ω. For example, if we take Ω to be the real line, $\mathscr{C}$ to be the system of open intervals and $\tau(\mathscr{C})$ to be the length of any interval C of $\mathscr{C}$, the resultant measure μ constructed from the pre-measure τ by Method I turns out to be usual Lebesgue measure (see §5 below).

***Theorem* 5.** *Let μ be a measure on Ω. Then μ is the measure, constructed by Method I, from the pre-measure τ, defined on the class of all sets of Ω to coincide with μ.*

Proof. Let $\mathscr{C}$ be the class of all subsets of Ω. Then τ is clearly a pre-measure. Let λ be the measure constructed from τ by Method I. Then, for all E of Ω, we have

$$\lambda(E) = \inf_{\cup C \supset E} \Sigma\tau(C_i) \leqslant \tau(E) = \mu(E),$$

as E lies in $\mathscr{C}$ and covers E. On the other hand, if $C_i \in \mathscr{C}$ and $\bigcup C \supset E$, on using the properties of μ as a measure

$$\sum_{i=1}^{\infty} \tau(C_i) = \sum_{i=1}^{\infty} \mu(C_i) \geqslant \mu\left(\bigcup_{i=1}^{\infty} C_i\right) \geqslant \mu(E),$$

so that
$$\lambda(E) = \inf_{\cup C \supset E} \Sigma\tau(C_i) \geqslant \mu(E).$$

Thus $\lambda = \mu$ as required.

2.6. We return to the relationship between a given measure μ defined on all the subsets of Ω and the countably additive measure ν defined on the σ-field of μ-measurable sets obtained by restricting μ to these sets. We prove a theorem which takes the first step towards the process of reconstructing μ when ν is given.

***Theorem* 6.** *Let ν be a countably additive measure defined on a σ-field $\mathscr{A}$ of sets. Then ν is a pre-measure. Let λ be the measure constructed from ν by Method I. Then the restriction of λ to the λ-measurable sets is an extension of ν.*

Proof. Clearly ν is a pre-measure. So the set function λ constructed from ν by Method I is a measure.

Consider any set E of Ω. Suppose $\{A_i\}$ is a sequence of sets of $\mathscr{A}$ with

$$E \subset \bigcup_{i=1}^{\infty} A_i.$$

Then, as $\mathscr{A}$ is a σ-field, the sets

$$A = \bigcup_{i=1}^{\infty} A_i,$$

$$B_i = A_i \backslash (\bigcup_{j<i} A_j) \quad (i = 1, 2, \ldots),$$

are in $\mathscr{A}$. As ν is a countably additive measure on $\mathscr{A}$, we have

$$\sum_{i=1}^{\infty} \nu(A_i) \geqslant \sum_{i=1}^{\infty} \nu(B_i) = \nu\left(\bigcup_{i=1}^{\infty} B_i\right) = \nu(A).$$

Hence $$\lambda(E) = \inf_{\cup A \supset E} \Sigma \nu(A_i) \geqslant \inf_{\substack{A \in \mathscr{A} \\ A \supset E}} \nu(A).$$

Taking any A in $\mathscr{A}$ with $A \supset E$ and considering the cover of E by the sequence $A, \varnothing, \varnothing, \ldots$ we have

$$\lambda(E) \leqslant \inf_{\substack{A \in \mathscr{A} \\ A \supset E}} \nu(A).$$

So $$\lambda(E) = \inf_{\substack{A \in \mathscr{A} \\ A \supset E}} \nu(A), \tag{16}$$

and we can choose a sequence $\{A_i\}$ of sets of $\mathscr{A}$ with

$$A_i \supset E, \quad \lambda(E) \leqslant \nu(A_i) \leqslant \lambda(E) + i^{-1}.$$

Then the set $A = \bigcap_{i=1}^{i=\infty} A_i$ belongs to $\mathscr{A}$ and satisfies

$$A \supset E, \quad \lambda(E) = \nu(A),$$

and the infimum in (16) is an attained infimum.

Now, when we take E to be a set B in $\mathscr{A}$, we have

$$\nu(B) = \inf_{\substack{A \in \mathscr{A} \\ A \supset B}} \nu(B) \leqslant \inf_{\substack{A \in \mathscr{A} \\ A \supset B}} \nu(A) \leqslant \nu(B),$$

so that, by (16),
$$\lambda(B) = \nu(B).$$

Thus ν is the restriction of λ to the σ-field $\mathscr{A}$.

It remains to prove that the sets of $\mathscr{A}$ are λ-measurable. Consider any set A of $\mathscr{A}$ and sets C, D separated by A with

$$C \subset A, \qquad D \subset \Omega \backslash A.$$

Then we can choose a set B in $\mathscr{A}$ with

$$C \cup D \subset B, \qquad \lambda(C \cup D) = \nu(B).$$

Now
$$C \subset A \cap B, \qquad D \subset (\Omega \backslash A) \cap B,$$

so that
$$\begin{aligned} \lambda(C) + \lambda(D) &\leqslant \nu(A \cap B) + \nu((\Omega \backslash A) \cap B) \\ &= \nu(\{A \cap B\} \cup \{(\Omega \backslash A) \cap B\}) \\ &= \nu(B) = \lambda(C \cup D). \end{aligned}$$

Hence each A in $\mathscr{A}$ is λ-measurable.

***Corollary* 1.** *We have*
$$\lambda(E) = \inf_{\substack{A \in \mathscr{A} \\ A \supset E}} \nu(A), \tag{17}$$
the infimum being attained.

If, in this theorem, we take ν to be the restriction of a measure μ to its class of measurable sets the measure λ will not in general coincide with the original measure μ. But this will happen if, and only if, μ is 'regular' in the following way.

***Definition* 6.** *A measure μ on Ω is said to be regular if for each E in Ω there is a μ-measurable set A with*

$$E \subset A, \qquad \mu(E) = \mu(A).$$

In terms of this definition we have a second corollary to Theorem 6.

***Corollary* 2.** *The measure λ is regular.*

Proof. As the sets of $\mathscr{A}$ are λ-measurable and, by Corollary 1, for each E in Ω there is a set A of $\mathscr{A}$ with

$$E \subset A, \qquad \lambda(E) = \nu(A) = \lambda(A),$$

the result follows.

We immediately obtain

Theorem 7. *Let μ be a measure on Ω. Let ν be the restriction of μ to the σ-field of μ-measurable sets. Then ν is a pre-measure, and the measure λ constructed from the pre-measure ν by Method I is a regular measure. All μ-measurable sets are λ-measurable and all λ-measurable sets of finite λ-measure are μ-measurable. Further λ coincides with μ, if, and only if, μ is regular.*

Proof. By theorem 3, the set function ν is a countably additive measure on the σ-field $\mathscr{M}$ of μ-measurable sets. By theorem 6, and its corollaries, the measure λ constructed from the pre-measure ν by Method I is a regular measure, coincides with μ and ν on $\mathscr{M}$ and satisfies

$$\lambda(E) = \inf_{\substack{A \in \mathscr{M} \\ A \supset E}} \mu(E),$$

for all E in Ω.

Further, still by theorem 6, all μ-measurable sets are λ-measurable. Suppose E is any λ-measurable set with finite λ-measure. By corollary 1 to theorem 6 there is a μ-measurable set E^* with

$$E \subset E^*, \qquad \lambda(E) = \mu(E^*).$$

Then

$$\lambda(E^*) = \mu(E^*) = \lambda(E).$$

As E is λ-measurable

$$\lambda(E) + \lambda(E^* \backslash E) = \lambda(E^*) = \lambda(E) < +\infty.$$

Hence

$$\lambda(E^* \backslash E) = 0.$$

So

$$\mu(E^* \backslash E) = 0.$$

Thus the null set $N = E^* \backslash E$ is μ-measurable. As E^* is μ-measurable it follows that

$$E = E^* \cap (\Omega \backslash N)$$

is μ-measurable, as required.

As λ is regular, λ and μ can only coincide, if μ is regular. If μ is regular, for all E in Ω we have

$$\inf_{\substack{A \in \mathscr{M} \\ A \supset E}} \mu(A) = \mu(E)$$

and λ coincides with μ.

Example. Let Ω be a space with cardinal $\aleph_2$. Let $|E|$ denote the cardinal of a set E, and define μ on the subsets of Ω by taking

$$\begin{aligned} \mu(E) &= 0, && \text{if } |E| \leqslant \aleph_0, \\ \mu(E) &= 1, && \text{if } \aleph_0 < |E| \leqslant \aleph_1, \\ \mu(E) &= +\infty, && \text{if } \aleph_1 < |E|. \end{aligned}$$

It is easy to verify that μ is a measure and that the μ-measurable sets are precisely the sets that are either countable or complements of countable sets.

The measure λ, defined by

$$\lambda(E) = \inf_{\substack{M \in \mathscr{M} \\ M \supset E}} \mu(M),$$

reduces to

$$\lambda(E) = 0, \quad \text{if } E \text{ is countable},$$

$$\lambda(E) = +\infty, \quad \text{if } E \text{ is uncountable}.$$

Thus all sets are λ-measurable.

The relationships between the pre-measure τ, the measure μ, the countably additive measure ν, that is also a pre-measure, and the regular λ can be illustrated, but not completely summarized by a diagram.

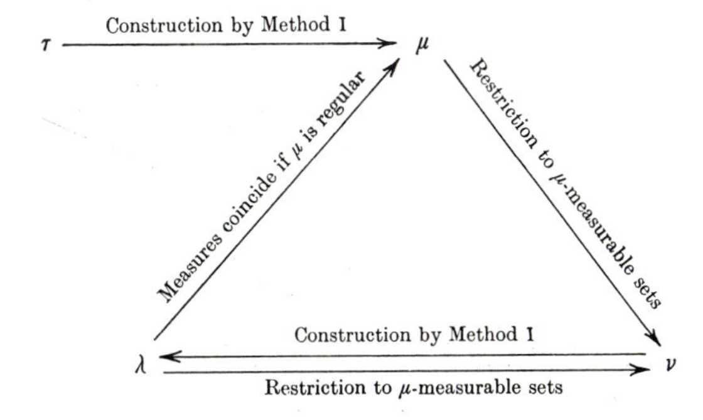

While our natural point of entry to this diagram is the pre-measure τ, from the point of view of some authors—for example, P. R. Halmos (1950)—the natural point of entry is the countably additive measure ν.

Later we shall need to work with a more general concept of regularity.

2.7. The next theorem gives two useful properties of measurable sets.

Theorem 8. *Let μ be a measure on Ω and let $\{A_i\}$ be a sequence of μ-measurable sets.*

(*a*) *If $A_1 \subset A_2 \subset A_3 \subset \dots$ and X is any set,*

$$\mu\left(X \cap \bigcap_{i=i}^{\infty} A_i\right) = \sup_i \mu(X \cap A_i).$$

(*b*) *If* $A_1 \supset A_2 \supset A_3 \supset \ldots$ *and* X *is any set and* $\mu(X \cap A_i)$ *is finite for some* i,

$$\mu\left(X \cap \bigcap_{i=1}^{\infty} A_i\right) = \inf_i \mu(X \cap A_i).$$

Proof. (*a*) Write

$$B_1 = A_1, \quad B_i = A_i \backslash A_{i-1} \quad (i = 2, 3, \ldots).$$

Then $\{B_i\}$ is a disjoint sequence of μ-measurable sets. By the corollary to theorem 2 (p. 8) we have

$$\begin{aligned}\mu\left(X \cap \bigcup_{i=1}^{\infty} A_i\right) &= \mu\left(X \cap \bigcup_{i=1}^{\infty} B_i\right)\\ &= \sum_{i=1}^{\infty} \mu(X \cap B_i)\\ &= \lim_{n\to\infty} \sum_{i=1}^{n} \mu(X \cap B_i)\\ &= \lim_{n\to\infty} \mu\left(X \cap \bigcup_{i=1}^{n} B_i\right)\\ &= \lim_{n\to\infty} \mu(X \cap A_n)\\ &= \sup_i \mu(X \cap A_i).\end{aligned}$$

(*b*) Suppose that $\mu(X \cap A_n)$ is finite. Write

$$B_i = A_n \backslash A_{n+i} \quad (i = 1, 2, \ldots).$$

Then $\{B_i\}$ is a sequence of μ-measurable sets with

$$B_1 \subset B_2 \subset B_3 \subset \ldots,$$

and by part (*a*), $$\mu\left(X \cap \bigcup_{i=1}^{\infty} B_i\right) = \sup_i \mu(X \cap B_i).$$

Thus $$\mu\left((X \cap A_n) \backslash \bigcap_{j=n+1}^{\infty} A_j\right) = \sup_i \mu(X \cap (A_n \backslash A_{n+i})). \tag{18}$$

As the sets $A_1, A_2, \ldots$ are μ-measurable

$$\mu(X \cap A_n) = \mu\left(X \cap \bigcap_{j=n+1}^{\infty} A_j\right) + \mu\left((X \cap A_n) \backslash \bigcap_{j=n+1}^{\infty} A_j\right),$$

$$\mu(X \cap A_n) = \mu(X \cap A_{n+i}) + \mu((X \cap A_n) \backslash A_{n+i}) \quad (i = 1, 2, \ldots).$$

As $\mu(X \cap A_n)$ is finite, so are the other measures in these formulae, and by (18)

$$\mu(X \cap A_n) - \mu\left(X \cap \bigcap_{j=n+1}^{\infty} A_j\right) = \sup_i \{\mu(X \cap A_n) - \mu(X \cap A_{n+i})\}.$$

Hence
$$\mu\left(X \cap \bigcap_{i=1}^{\infty} A_i\right) = \mu\left(X \cap \bigcap_{j=n+1}^{\infty} A_j\right)$$
$$= \inf_i \mu(X \cap A_{n+i})$$
$$= \inf_i \mu(X \cap A_i)$$
as required.

2.8. The next theorem shows that, when the measure is regular, the first part of the last theorem extends to any increasing sequence of sets.

***Theorem* 9.** *Let μ be a regular measure on Ω. Let $\{A_i\}$ be any increasing sequence of sets of Ω. Then*

$$\mu\left(\bigcup_{i=1}^{\infty} A_i\right) = \sup_i \mu(A_i). \tag{19}$$

Proof. Trivially

$$\sup_i \mu(A_i) \leqslant \sup_i \mu\left(\bigcup_{j=1}^{\infty} A_j\right) = \mu\left(\bigcup_{i=1}^{\infty} A_i\right). \tag{20}$$

Since μ is regular, for each i we can choose a μ-measurable set B_i with
$$A_i \subset B_i, \qquad \mu(A_i) = \mu(B_i).$$

Write
$$C_i = \bigcap_{j \geqslant i} B_j \quad (i = 1, 2, \ldots).$$

Then, as
$$A_i \subset A_j \subset B_j \quad (\text{for } j \geqslant i),$$

we have
$$A_i \subset C_i \subset B_i$$

and
$$\mu(A_i) \leqslant \mu(C_i) \leqslant \mu(B_i) = \mu(A_i)$$

for $i = 1, 2, \ldots$. Further $\{C_i\}$ is an increasing sequence of μ-measurable sets. So, by theorem 8, part (*a*), we have

$$\mu\left(\bigcup_{i=1}^{\infty} A_i\right) \leqslant \mu\left(\bigcup_{i=1}^{\infty} C_i\right) = \sup_i \mu(C_i) = \sup_i \mu(A_i).$$

Hence, by (20), we have (19) as required.

Remark. While theorem 9 shows that when μ is regular, the first part of theorem 8 holds for an arbitrary sequence of increasing sets, there is no corresponding extension of the second part of theorem 8 to an arbitrary sequence of decreasing sets. One simple example to show this, is obtained by taking μ to be the measure defined on a space X with at least a countable sequence $\{x_j\}$ of distinct points by defining

$$\mu(\varnothing) = 0,$$
$$\mu(E) = 1 \quad (\text{if } E \neq \varnothing).$$

It is easy to verify that μ is a measure, that $\varnothing$ and X are the only μ-measurable sets and that μ is a regular measure. If we then take A_i to be the set of all points x_j with $j \geqslant i$, we have $A_1 \supset A_2 \supset A_3 \supset \ldots$,

$$\mu(A_i) = 1 \quad (i \geqslant 1),$$

but
$$\mu\left(\bigcap_{i=1}^{\infty} A_i\right) = \mu(\varnothing) = 0 < 1 = \inf_i \mu(A_i).$$

We give a second example to show that the result breaks down for Lebesgue measure (see §5 below). In this paragraph we use λ to denote Lebesgue measure on the real line. We use the axiom of choice to construct certain non-measurable sets. We introduce an equivalence relation regarding two real numbers as equivalent if their difference is a rational number. Each equivalence class of real numbers has many elements in the closed interval $I = [0, 1]$. Form a set N by choosing just one representative in I from each equivalence class of real numbers. Let $r_1, r_2, \ldots$ be an enumeration of the rational numbers, and let $N(r_i)$ denote the translate of the set N in the positive direction by the rational displacement r_i, $i = 1, 2, \ldots$. Then the sets $N(r_i)$, $i = 1, 2, \ldots$ are all disjoint, and their union is the whole real line. By the fact that Lebesgue measure has a translation invariant definition we have

$$\lambda(N(r_i)) = \lambda(N) \leqslant \lambda(I) = 1$$

for all i. If we had $\lambda(N) = 0$, we would have

$$1 = \lambda(I) \leqslant \lambda\left(\bigcup_{i=1}^{\infty} N(r_i)\right)$$
$$\leqslant \sum_{i=1}^{\infty} \lambda(N(r_i)) = 0.$$

Hence
$$0 < \lambda(N) \leqslant 1.$$

Our definition of N is in fact not sufficiently precise to enable us to determine the exact value of $\lambda(N)$. Given any α, β with $0 < \alpha < \beta < 1$, it might happen that our choice of the points to place in N always leads to a point in the interval $[\alpha, \beta]$ in which case we would have $\lambda(N) \leqslant \beta - \alpha$. So although we know that $\lambda(N)$ is positive, we cannot name a positive lower bound for this number. But to return from this digression to the point of our example, we take the decreasing sequence of sets to be the sets

$$A_i = \bigcup_{j \geqslant i} N(r_j) \cap J,$$

where J denotes the interval $[0, 2]$. Then each set A_i is contained in J, but contains one of the sets $N(r_j)$. Thus $0 < \lambda(N) < \lambda(A_i) \leqslant 2$, for $i = 1, 2, \ldots$; but

$$\lambda\left(\bigcap_{i=1}^{\infty} A_i\right) = \lambda(\varnothing) = 0.$$

So

$$\lambda\left(\bigcap_{i=1}^{\infty} A_i\right) = 0 < \lambda(N) \leqslant \inf_i \lambda(A_i),$$

and (*b*) of theorem 8 fails. We draw from this failure the conclusion that the set N is not Lebesgue measurable.

2.9. The regular measures have a further advantage. The measurability criterion for sets of finite measure can be put in a simple form.

***Theorem* 10.** *Let μ be a regular measure. Let M be a μ-measurable subset of Ω with $\mu(M)$ finite. A subset E of M is μ-measurable, if, and only if,*

$$\mu(M) = \mu(E) + \mu(M \backslash E). \tag{21}$$

Proof. If E is μ-measurable, E and $M \backslash E$ are disjoint μ-measurable sets so that (21) holds.

Now suppose that E is a subset of M satisfying (21). Write

$$F = M \backslash E.$$

As μ is regular we can choose μ-measurable sets E^*, F^* with

$$E \subset E^* \subset M, \qquad \mu(E) = \mu(E^*),$$

$$F \subset F^* \subset M, \qquad \mu(F) = \mu(F^*).$$

Using the measurability of E^*, F^*, we have

$$\begin{aligned}\mu(E^* \backslash F^*) + \mu(E^* \cap F^*) + \mu(F^* \backslash E^*) &= \mu(E^* \cup F^*) \\ &\geqslant \mu(E \cup F) = \mu(M).\end{aligned} \tag{22}$$

Similarly, using also (21), we obtain

$$\begin{aligned}\mu(E^*\backslash F^*)+2\mu(E^*\cap F^*)+\mu(F^*\backslash E^*) &= \mu(E^*)+\mu(F^*)\\ &= \mu(E)+\mu(F)\\ &= \mu(E)+\mu(M\backslash E)\\ &= \mu(M) < +\infty. \qquad (23)\end{aligned}$$

Hence comparing (22) and (23), it follows that

$$\mu(E^*\cap F^*) = 0.$$

Now, writing $N = E^*\backslash E$ we have

$$N = E^*\backslash E = E^*\cap(M\backslash E) = E^*\cap F \subset E^*\cap F^*,$$

so that $\mu(N) = 0$. Thus

$$E = E^*\cap(\Omega\backslash N),$$

where E^* and N are μ-measurable. Hence E is μ-measurable as required.

2.10. Our next two theorems will give methods of obtaining new measures from a given measure or a given family of measures.

***Theorem* 11.** *Let μ be a measure on a space Ω and let X be any set of Ω. Then the set function μ_X, defined by*

$$\mu_X(E) = \mu(E\cap X),$$

is a measure on Ω. The μ-measurable sets are all μ_X-measurable.

Proof. (*a*) Clearly $0 \leqslant \mu_X(E) \leqslant +\infty$ for all E in Ω.

(*b*) Also $\mu_X(\varnothing) = \mu(\varnothing) = 0$.

(*c*) Further, if $E_1 \subset E_2$, then

$$\mu_X(E_1) = \mu(E_1\cap X) \leqslant \mu(E_2\cap X) = \mu_X(E_2).$$

(*d*) If $\{E_i\}$ is any sequence of sets of Ω, then

$$\mu_X\left(\bigcup_{i=1}^{\infty} E_i\right) = \mu\left(\bigcup_{i=1}^{\infty}(E_i\cap X)\right) \leqslant \sum_{i=1}^{\infty}\mu(E_i\cap X) = \sum_{i=1}^{\infty}\mu_X(E_i).$$

(*e*) If E is any μ-measurable set and A and B are separated by E, then $A\cap X$ and $B\cap X$ are separated by E, so that

$$\begin{aligned}\mu_X(A\cup B) &= \mu((A\cap X)\cup(B\cap X))\\ &= \mu(A\cap X)+\mu(B\cap X)\\ &= \mu_X(A)+\mu_X(B).\end{aligned}$$

The required results now follow immediately from (*a*) to (*e*).

2.11. We conclude this section with a theorem showing how the supremum of a family of measures is a measure.

***Theorem* 12.** *Let I be an arbitrary index set. Suppose that for all ι in I, μ_ι is a measure on Ω. Then*

$$\mu(E) = \sup_{\iota \in I} \mu_\iota(E)$$

is a measure on Ω.

Proof. (*a*) Clearly $0 \leqslant \mu(E) \leqslant +\infty$ for all in Ω.

(*b*)

$$\mu(\varnothing) = \sup_{\iota \in I} \mu_\iota(\varnothing) = 0.$$

(*c*) If $E_1 \subset E_2$ then

$$\mu(E_1) = \sup_{\iota \in I} \mu_\iota(E_1) \leqslant \sup_{\iota \in I} \mu_\iota(E_2) = \mu(E_2).$$

(*d*) Suppose $\{E_j\}$ is any sequence of sets of Ω. Then, for all ι of I, we have

$$\mu_\iota\left(\bigcup_{j=1}^{\infty} E_j\right) \leqslant \sum_{j=1}^{\infty} \mu_\iota(E_j) \leqslant \sum_{j=1}^{\infty} \mu(E_j),$$

so that

$$\mu\left(\bigcup_{j=1}^{\infty} E_j\right) \leqslant \sum_{j=1}^{\infty} \mu(E_j),$$

as required.

We have now verified that μ is a measure.

Corollary. *Let I be an arbitrary index set. Suppose that for all ι in I, μ_ι is a measure on Ω. Then there is a unique measure μ on Ω with the properties*:

(*a*) *for each E of Ω* $\quad \mu(E) \leqslant \inf_{\iota \in I} \mu_\iota(E)$,

(*b*) *for each measure ν on Ω with*

$$\nu(E) \leqslant \inf_{\iota \in I} \mu_\iota(E)$$

for all E of Ω, we have $\quad \mu(E) \geqslant \nu(E)$,
for all E of Ω.

Proof. It suffices to take

$$\mu(E) = \sup_{\nu} \nu(E),$$

the supremum being over all measures ν on Ω with the property

$$\nu(E) \leqslant \inf_{\iota \in I} \mu_\iota(E),$$

for all E of Ω. The set of such measures ν is always non-empty as it always contains the measure that assigns the value 0 to each set.

Remark. The class of all measures is partially ordered by the relation '$\mu \leqslant \nu$, if, and only if $\mu(E) \leqslant \nu(E)$ for all E of Ω'. The theorem and its corollary show that the measures form a complete lattice under this partial order.

§ 3 Measures in topological spaces

In this section we first introduce the idea of a topological space. Although topological ideas do not enter the formulations of the main results of this section they are needed to point and motivate them. We first show that, if a measure μ is constructed by Method I from a pre-measure τ defined on a class $\mathscr{C}$ of sets with $\Omega \in \mathscr{C}$, then μ is $\mathscr{C}_{\sigma\delta}$-regular. Then we show that, if a measure μ is $\mathscr{R}$-regular, each μ-measurable set with a finite μ-measure contains a set, with the same μ-measure, that is a difference between sets of $\mathscr{R}$.

3.1. Although there are various alternative ways of introducing topological ideas into a space we will work with the concept of an open set as the primitive concept. The standard axioms for the open sets of a space Ω are:

(*a*) $\varnothing$ and Ω are open;

(*b*) the intersection of two open sets is open;

(*c*) the union of any collection of open sets is open.

***Definition* 7.** *A space Ω is said to be a topological space as soon as a system of 'open' sets has been specified so that they satisfy the conditions* (*a*), (*b*) *and* (*c*) *above. The system of open sets is called the topology.*

After the open sets and the closed sets—that is, the complements of open sets—the Borel sets will be the most significant class of sets in a topological space.

***Definition* 8.** *The Borel sets of a topological space Ω are the sets of the minimal σ-field of sets of Ω containing the open sets of Ω.*

This definition begs the question of the existence of such a minimal σ-field; before we can use it (and strictly speaking before we write it down) we need the following lemma of justification.

***Lemma* 1.** *If $\mathscr{H}$ is any class of sets of Ω there is a unique minimal σ-field containing $\mathscr{H}$.*

Proof. First note that the class $\mathscr{S}$ of all subsets of Ω is a σ-field containing $\mathscr{H}$.

Let $\mathbf{F}$ be the super-system of all σ-fields $\mathscr{F}$ of sets of Ω containing $\mathscr{H}$. Then $\mathbf{F}$ contains $\mathscr{S}$ and so is not empty. Write

$$\mathscr{K} = \bigcap_{\mathscr{F} \in \mathbf{F}} \mathscr{F}. \tag{1}$$

Now, if H is any set of $\mathscr{H}$, then

$$H \in \mathscr{H} \subset \mathscr{F}$$

for all $\mathscr{F}$ of $\mathbf{F}$. Hence $\mathscr{H} \subset \mathscr{K}$.

Now clearly $\varnothing \in \mathscr{K}$ as $\varnothing$ belongs to every σ-field.

Further, if $K \in \mathscr{K}$, then for all $\mathscr{F}$ of $\mathbf{F}$ we have $K \in \mathscr{F}$, so that $\Omega \backslash K \in \mathscr{F}$. Thus, if $K \in \mathscr{K}$, then

$$\Omega \backslash K \in \bigcap_{\mathscr{F} \in \mathbf{F}} \mathscr{F} = \mathscr{K}.$$

Finally, if $\{K_i\}$ is a sequence of sets of $\mathscr{K}$, then for each $\mathscr{F}$ of $\mathbf{F}$, $\{K_i\}$ is a sequence of sets of $\mathscr{F}$, so that $\bigcup_{i=1}^{i=\infty} K_i \in \mathscr{F}$. Thus, if $\{K_i\}$ is a sequence of sets of $\mathscr{K}$, then $\bigcup_{i=1}^{i=\infty} K_i \in \mathscr{K}$.

The last four paragraphs show that $\mathscr{K}$ is a σ-field containing $\mathscr{H}$. By the definition (1), the σ-field $\mathscr{K}$ is the unique minimal σ-field containing $\mathscr{H}$, as it is the intersection of all the σ-fields containing $\mathscr{H}$.

3.2. We now turn to measure-theoretic considerations. We generalize the idea of a regular measure; see definition 6 above.

***Definition* 9.** *If $\mathscr{R}$ is a class of sets, a measure μ is said to be $\mathscr{R}$-regular, if for each E in Ω there is a set R in $\mathscr{R}$ with*

$$E \subset R \quad \textit{and} \quad \mu(E) = \mu(R).$$

Then, comparing this definition with definition 6, we see that μ is regular, if μ is $\mathscr{M}$-regular, $\mathscr{M}$ being the class of μ-measurable sets.

Although this definition of $\mathscr{R}$-regularity is given in terms of an arbitrary class $\mathscr{R}$ of sets, it is only of significance, as far as I know, when $\mathscr{R}$ is a class of 'topologically respectable' sets or the class $\mathscr{M}$ of μ-measurable sets.

Our next theorem will show that, if a measure μ has been constructed by Method I from a pre-measure defined only on a 'topologically respectable' class of sets, we have an automatic method of deducing that μ is $\mathscr{R}$-regular for a second 'topologically respectable' class $\mathscr{R}$. To state the result in a concise form it is convenient to introduce a definition.

***Definition* 10.** *If $\mathscr{R}$ is a class of sets the classes $\mathscr{R}_\sigma$ and $\mathscr{R}_\delta$ are defined, by taking $\mathscr{R}_\sigma$ to be the class of all countable set unions*

$$\bigcup_{i=1}^{\infty} R_i \quad \textit{with} \quad R_i \in \mathscr{R},$$

and by taking $\mathscr{R}_\delta$ to be the class of all countable intersections

$$\bigcap_{i=1}^{\infty} R_i \quad \textit{with} \quad R_i \in \mathscr{R}.$$

***Theorem* 13.** *Let μ be the measure constructed by Method I from a pre-measure τ defined on a class $\mathscr{C}$ of sets, with $\Omega \in \mathscr{C}$. Then μ is $\mathscr{C}_{\sigma\delta}$-regular.*

Proof. Let E be any set in Ω. If $\mu(E) = +\infty$, we have

$$E \subset \Omega, \qquad \mu(E) = \mu(\Omega),$$

and Ω being in $\mathscr{C}$ is in $\mathscr{C}_{\sigma\delta}$. Thus we can confine our attention to a set E in Ω with $\mu(E) < +\infty$. As

$$\mu(E) = \inf_{\cup C \supset E} \Sigma\tau(C_i),$$

for each integer $j \geqslant 1$ we can choose a sequence $\{C_i^{(j)}\}_{i=1}^{i=\infty}$ of sets of $\mathscr{C}$ with

$$\bigcup_{i=1}^{\infty} C_i^{(j)} \supset E \quad \text{and} \quad \sum_{i=1}^{\infty} \tau(C_i^{(j)}) < \mu(E) + j^{-1}.$$

Write

$$D = \bigcap_{j=1}^{\infty} \bigcup_{i=1}^{\infty} C_i^{(j)}.$$

Then clearly D is in $\mathscr{C}_{\sigma\delta}$ and $E \subset D$. Further, for each $j \geqslant 1$ we have

$$\mu(E) \leqslant \mu(D) = \inf_{\cup C \supset D} \Sigma\tau(C_i)$$

$$\leqslant \sum_{i=1}^{\infty} \tau(C_i^{(j)}) < \mu(E) + j^{-1},$$

since $\{C_i^{(j)}\}_{i=1}^{i=\infty}$ is a sequence of sets of $\mathscr{C}$ covering D. Hence

$$\mu(E) = \mu(D).$$

This shows that μ is $\mathscr{C}_{\sigma\delta}$-regular, as required.

***Corollary*.** *If τ is defined only on open sets, then μ is $\mathscr{G}_\delta$-regular. If τ is defined only on Borel sets, then μ is Borel-regular.*

Proof. If τ is not already defined on the set Ω we can extend its definition by writing $\tau(\Omega) = +\infty$ without changing μ. Then the results follow on noting that $\mathscr{G}_\sigma = \mathscr{G}$ and that the class of Borel sets is closed under countable unions and under countable intersections.

3.3. The concept of regularity is concerned with the approximation of sets in measure from without by sets of some specified class. We are sometimes concerned with corresponding problems of approximation from within. There are two limitations that arise when we consider such approximations. In the first place we can only hope to get good approximations from within when the given set is measurable. Secondly, in the abstract theory, we have to confine our attention to sets of finite measure. We obtain:

Theorem 14. *Let $\mathscr{R}$ be a system of sets of Ω and let μ be an $\mathscr{R}$-regular measure. Let E be a μ-measurable set with finite μ-measure. Then there is a set C, of the form $R_1 \backslash R_2$ with R_1 and R_2 in $\mathscr{R}$, such that*

$$C \subset E, \qquad \mu(C) = \mu(E).$$

Proof. As μ is $\mathscr{R}$-regular, we can choose a set R_1 of $\mathscr{R}$ with

$$E \subset R_1, \qquad \mu(E) = \mu(R_1).$$

As E is μ-measurable,

$$\mu(E) = \mu(R_1) = \mu(R_1 \cap E) + \mu(R_1 \backslash E)$$
$$= \mu(E) + \mu(R_1 \backslash E).$$

As $\mu(E)$ is finite, this implies that $\mu(R_1 \backslash E) = 0$. As μ is $\mathscr{R}$-regular we can choose R_2 in $\mathscr{R}$ with

$$R_1 \backslash E \subset R_2, \qquad 0 = \mu(R_1 \backslash E) = \mu(R_2).$$

Write $C = R_1 \backslash R_2$. Then

$$C = R_1 \backslash R_2 \subset R_1 \backslash \{R_1 \backslash E\} = E.$$

Further, as $\mu(R_2) = 0$,

$$\begin{aligned} \mu(C) &= \mu(R_1 \backslash R_2) \\ &= \mu(R_1 \backslash R_2) + \mu(R_2) \\ &\geqslant \mu(\{R_1 \backslash R_2\} \cup R_2) \\ &\geqslant \mu(R_1) = \mu(E) \geqslant \mu(C). \end{aligned}$$

So
$$C \subset E, \qquad \mu(C) = \mu(E),$$
as required.

Corollary. *The sets R_1, R_2 can be chosen with*

$$\mu(R_1) = \mu(E), \qquad \mu(R_2) = 0.$$

3.4. While we have proved quite a number of theorems about measures in abstract spaces and in topological spaces the theorems have hardly started to fit together to form a theory. Despite the examples listed in Chapter 1, §2 we have rather few significant examples available for study. Further we have no effective way of showing that we have reasonably large classes of measurable sets. The limitations will be removed in the next section where we confine our attention to metric spaces.

§4 Measures in metric spaces

In this section we define the concept of a metric and relate it to the concept of a topology. We then introduce a second method, 'Method II', of obtaining a measure from a pre-measure. The names 'Method I' and 'Method II' are due to Munroe (1953). This second method has the advantage that it always leads to a measure for which the Borel sets are measurable. We prove this in two stages, first showing that any Method II measure is what we call a 'metric measure', being additive over any pair of disjoint sets that are separated from each other by a positive distance, and then showing that Borel sets are measurable with respect to any metric measure. We then investigate the regularity and approximation properties of measures constructed by Method II.

4.1. A metric is a function on a space that generalizes the elementary notion of the distance between two points.

***Definition* 11.** *A function $\rho(x,y)$ defined for all pairs of points x, y of a space Ω is called a metric if, for all points x, y, z of Ω:*

(*a*) $\rho(x,y) \geqslant 0$, *and* $\rho(x,y) = 0$, *if, and only if*, $x = y$;

(*b*) $\rho(x,y) = \rho(y,x)$;

(*c*) $\rho(x,z) \leqslant \rho(x,y)+\rho(y,z)$.

A space is called a metric space, if it is an abstract space, on which a metric is defined. A topology, called the metric topology, can be introduced into any metric space by defining the open sets of the metric space to be the sets G with the property:

> if x is a point of G, then for some $\delta > 0$, all points y with $\rho(x,y) < \delta$ also belong to G.

It is easy to verify that the open sets defined in this way satisfy the conditions (*a*), (*b*), (*c*) stated at the beginning of §3, so that the metric topology is necessarily a topology.

We introduce the concept of the diameter of a set.

***Definition* 12.** *The diameter of a set E of a metric space Ω with metric ρ is denoted by $d(E)$ and defined by*

$$d(E) = \sup_{x, y \in E} \rho(x, y),$$

with the convention $$d(\varnothing) = 0.$$

When we come to study Hausdorff measures we shall find that the diameter is indispensible for defining the pre-measures that lead to the Hausdorff measures. In the meantime we shall use the diameter in quite a different way. We introduce the second method of constructing a measure from a given pre-measure.

***Theorem* 15.** *If τ is a pre-measure defined on a class $\mathscr{C}$ of sets, in a metric space Ω with metric ρ, the set function*

$$\mu(E) = \sup_{\delta>0} \mu_\delta(E), \tag{1}$$

where $$\mu_\delta(E) = \inf_{\substack{C_i \in \mathscr{C},\, d(C_i) \leqslant \delta \\ \cup C_i \supset E}} \sum_{i=1}^{\infty} \tau(C_i), \tag{2}$$

is a measure on Ω.

Remarks. When no confusion can arise we will abbreviate the formula (2) to the form

$$\mu_\delta(E) = \inf_{\substack{d(C) \leqslant \delta \\ \cup C \supset E}} \Sigma \tau(C_i). \tag{3}$$

We shall call the measure μ the measure constructed from the pre-measure τ by Method II. As the effect of reducing δ is to reduce the class of covers over which the infimum (2) is taken, $\mu_\delta(E)$ does not decrease as δ decreases, and it is the small values of δ that are relevant in taking the supremum (1); indeed we could immediately replace the formula (1) by

$$\mu(E) = \lim_{\delta \to 0+} \mu_\delta(E). \tag{4}$$

In other words it is the 'fine' covers, i.e. those by sets of small diameter, that determine $\mu(E)$.

Proof. For each δ with $\delta > 0$ let $\mathscr{C}_\delta$ be the system of all sets C of $\mathscr{C}$ with $d(C) \leqslant \delta$ (here, of course, δ is a numerical suffix, not an operator indicating the taking of countable intersections). Then, it follows immediately that the restriction τ_δ of τ to $\mathscr{C}_\delta$ is a pre-measure on Ω. Then, for each δ with $\delta > 0$, we have

$$\mu_\delta(E) = \inf_{\substack{C_i \in \mathscr{C},\, d(C_i) \leqslant \delta \\ \cup C_i \supset E}} \sum_{i=1}^{\infty} \tau(C_i) = \inf_{\substack{C_i \in \mathscr{C}_\delta \\ \cup C_i \supset E}} \sum_{i=1}^{\infty} \tau_\delta(C_i),$$

and $\mu_\delta(E)$ is the measure constructed by Method I from the pre-measure τ_δ. By theorem 4 this set function $\mu_\delta(E)$ is a measure. By theorem 12 the set function

$$\mu(E) = \sup_{\delta>0} \mu_\delta(E)$$

is a measure, as required.

4.2. The main advantages of the measures constructed by Method II over those constructed by Method I all stem from the next theorem showing that the Method II measures are additive when considered on the union of a pair of sets that are at a 'positive distance apart'. To make this idea precise we introduce the

***Definition* 13.** *If A and B are disjoint non-empty sets in a metric space Ω with metric ρ, A and B are said to be positively separated if the distance*

$$\inf_{a\in A,\ b\in B} \rho(a,b)$$

separating A and B is positive.

***Theorem* 16.** *Let μ be a measure on a metric space Ω constructed from a pre-measure τ by Method II. If A and B are disjoint non-empty sets of Ω that are positively separated, then*

$$\mu(A \cup B) = \mu(A) + \mu(B).$$

Proof. As μ is a measure

$$\mu(A \cup B) \leqslant \mu(A) + \mu(B).$$

So we need to prove that

$$\mu(A \cup B) \geqslant \mu(A) + \mu(B).$$

In proving this we may clearly suppose $\mu(A \cup B) < +\infty$. The aim will be to obtain, from a fine economical cover of $A \cup B$, separate fine economical covers that cover A and B separately by a 'filtering out' process.

As A and B are disjoint non-empty and positively separated we can choose $\delta > 0$ so small that

$$\inf_{a\in A,\ b\in B} \rho(a,b) \geqslant \delta.$$

Let $\epsilon > 0$ be given. Let δ_1, δ_2 be given with $0 < \delta_1 < d(\Omega)$,

$$0 < \delta_2 < d(\Omega).$$

Write $$\eta = \min\{\delta_1, \delta_2, \tfrac{1}{2}\delta\}.$$

Since, with the usual notation,

$$\mu(A \cup B) = \sup_{d>0} \inf_{\substack{d(C) \leqslant d \\ \cup C \supset A \cup B}} \Sigma\tau(C_i),$$

and $\mu(A \cup B) < +\infty$, we have

$$\inf_{d(C) \leqslant \eta,\, \cup C \supset A \cup B} \Sigma\tau(C_i) \leqslant (A \cup B),$$

and we can choose a sequence $\{C_i\}$ of sets of $\mathscr{C}$ with

$$d(C_i) \leqslant \eta \quad (i = 1, 2, \ldots),$$

$$\bigcup_{i=1}^{\infty} C_i \supset A \cup B,$$

$$\sum_{i=1}^{\infty} \tau(C_i) \leqslant \mu(A \cup B) + \epsilon.$$

Now, if for some i we had both

$$C_i \cap A \neq \varnothing, \qquad C_i \cap B \neq \varnothing,$$

there would be points a_0, b_0 with

$$a_0 \in C_i \cap A, \qquad b_0 \in C_i \cap B$$

so that

$$\rho(a_0, b_0) \leqslant d(C_i) \leqslant \eta \leqslant \tfrac{1}{2}\delta \leqslant \tfrac{1}{2} \inf_{a \in A,\, b \in B} \rho(a, b) \leqslant \tfrac{1}{2}\rho(a_0, b_0),$$

with the contradictory implication $\rho(a_0, b_0) = 0$.

So, for no i do we have both

$$C_i \cap A \neq \varnothing \quad \text{and} \quad C_i \cap B \neq \varnothing. \tag{5}$$

For $i = 1, 2, \ldots$, write

$$A_i = C_i, \quad \text{if} \quad C_i \cap A \neq \varnothing,$$

$$A_i = \varnothing, \quad \text{if} \quad C_i \cap A = \varnothing,$$

$$B_i = C_i, \quad \text{if} \quad C_i \cap B \neq \varnothing,$$

$$B_i = \varnothing, \quad \text{if} \quad C_i \cap B = \varnothing.$$

Then, clearly $\displaystyle\bigcup_{i=1}^{\infty} A_i \supset \bigcup_{i=1}^{\infty} \{C_i \cap A\} = \left\{\bigcup_{i=1}^{\infty} C_i\right\} \cap A = A,$

$$\bigcup_{i=1}^{\infty} B_i \supset \bigcup_{i=1}^{\infty} \{C_i \cap B\} = \left\{\bigcup_{i=1}^{\infty} C_i\right\} \cap B = B.$$

Further each sequence $\{A_i\}$ and $\{B_i\}$ is a sequence of sets of $\mathscr{C}$ of diameter not more than δ. As the possibility (5) does not arise we have either

$$\tau(A_i)+\tau(B_i) = \tau(\varnothing)+\tau(\varnothing) = 0,$$

or

$$\tau(A_i)+\tau(B_i) = \tau(C_i)+\tau(\varnothing) = \tau(C_i);$$

in either case

$$\tau(A_i)+\tau(B_i) \leqslant \tau(C_i).$$

Thus

$$\sum_{i=1}^{\infty}\tau(A_i)+\sum_{i=1}^{\infty}\tau(B_i) \leqslant \sum_{i=1}^{\infty}\tau(C_i) \leqslant \mu(A \cup B)+\epsilon.$$

Now

$$\{A_i\} \subset \mathscr{C}, \qquad d(A_i) \leqslant \eta \leqslant \delta_1, \qquad \bigcup_{i=1}^{\infty} A_i \supset A,$$

$$\{B_i\} \subset \mathscr{C}, \qquad d(B_i) \leqslant \eta \leqslant \delta_2, \qquad \bigcup_{i=1}^{\infty} B_i \supset B.$$

Hence

$$\mu_{\delta_1}(A) \leqslant \sum_{i=1}^{\infty}\tau(A_i),$$

$$\mu_{\delta_2}(B) \leqslant \sum_{i=1}^{\infty}\tau(B_i),$$

and

$$\mu_{\delta_1}(A)+\mu_{\delta_2}(B) \leqslant \mu(A \cup B)+\epsilon.$$

As this holds for all pairs δ_1, δ_2 with $0 < \delta_1 < d(\Omega)$, $0 < \delta_2 < d(\Omega)$, it follows that

$$\mu(A)+\mu(B) \leqslant \mu(A \cup B)+\epsilon.$$

As this holds for all $\epsilon > 0$ we have

$$\mu(A)+\mu(B) \leqslant \mu(A \cup B),$$

as required.

4.3. This last property of the measures constructed by Method II is so important that it is used as a definition.

***Definition* 14.** *A measure μ defined on a metric space Ω is called a metric measure, if*

$$\mu(A \cup B) = \mu(A)+\mu(B),$$

for every pair of disjoint non-empty sets A, B that are positively separated.

So, in terms of this definition, theorem 16 asserts that *a measure in a metric space constructed by Method II is always a metric measure.*

Our next result is a result on what might be called 'strictly increasing' sequences of sets. It should be compared with theorems 8 and 9; it is a form of 'Carathéodory's Lemma'.

***Theorem* 17.** *Let μ be a metric measure on a metric space Ω. Let $A_1, A_2, \ldots,$ be a sequence of sets with*

$$A_1 \subset A_2 \subset A_3 \subset \ldots,$$

and suppose that, for each $n \geqslant 1$, the sets A_n and $\Omega \backslash A_{n+1}$ are positively separated. Then

$$\mu\left(\bigcup_{n=1}^{\infty} A_n\right) = \sup_n \mu(A_n).$$

Proof. Write
$$A = \bigcup_{n=1}^{\infty} A_n.$$

Trivially
$$\mu(A) \geqslant \sup_n \mu(A_n).$$

So it suffices to prove that

$$\sup_n \mu(A_n) \geqslant \mu(A).$$

Thus we may suppose that $\sup_n \mu(A_n)$ is finite.

Write $D_1 = A_1$ and write $D_n = A_n \backslash A_{n-1}$ for $n \geqslant 2$. If m, n are integers with $n \geqslant 1$ and $m \geqslant n+2$, we have

$$D_n \subset A_n$$

$$D_m \subset \Omega \backslash A_{m-1} \subset \Omega \backslash A_{n+1}.$$

As A_n and $\Omega \backslash A_{n+1}$ are positively separated, it follows that D_n and D_m are positively separated when $n \geqslant 1$ and $m \geqslant n+2$. Further the sets

$$\bigcup_{k=1}^{n} D_{2k+r}, \qquad D_{2n+2+r},$$

will be positively separated for $r = 0$ or -1 and $n \geqslant 1$. As μ is a metric measure, it follows, by inductive use of theorem 16, that

$$\mu\left(\bigcup_{k=1}^{n} D_{2k+r}\right) = \sum_{k=1}^{n} \mu(D_{2k+r}),$$

for $r = 0$ or -1 and $n \geqslant 1$.

Now
$$A_{2n} = \left[\bigcup_{k=1}^{n} D_{2k}\right] \cup \left[\bigcup_{k=1}^{n} D_{2k-1}\right],$$

so that

$$\sum_{k=1}^{n} \mu(D_{2k+r}) = \mu\left(\bigcup_{k=1}^{n} D_{2k+r}\right) \leqslant \mu(A_{2n}) \leqslant \sup_m \mu(A_m) < +\infty,$$

for $r = 0, -1$. Hence the two series

$$\sum_{k=1}^{\infty} \mu(D_{2k}), \qquad \sum_{k=1}^{\infty} \mu(D_{2k-1}) \tag{6}$$

converge.

Now
$$\begin{aligned}\mu(A) &= \mu\left(\bigcup_{n=1}^{\infty} A_n\right) \\ &= \mu\left(A_n \cup \left\{\bigcup_{k=n+1}^{\infty} D_k\right\}\right) \\ &\leqslant \mu(A_n) + \sum_{k=n+1}^{\infty} \mu(D_k) \\ &\leqslant \{\sup_m \mu(A_m)\} + \sum_{k=n+1}^{\infty} \mu(D_k). \end{aligned} \tag{7}$$

Since both the series (6) converge, we may let n tend to infinity in (7) and obtain
$$\mu(A) \leqslant \sup_m \mu(A_m),$$
as required.

***Theorem* 18.** *If μ is a metric measure in a metric space Ω all closed sets of Ω are μ-measurable.*

Proof. Let F be a closed set in Ω. Consider any two sets A, B that are separated by F with

$$A \subset F, \qquad B \subset \Omega \backslash F.$$

We may suppose that A, B are both non-empty. Our aim is to express B as a union of an increasing sequence

$$B_1 \subset B_2 \subset B_3 \subset \dots$$

of sets with B_n and $\Omega \backslash B_{n+1}$ positively separated for each $n \geqslant 1$ and with A and B_n positively separated for each $n \geqslant 1$.

Once we can effect such a construction it will follow, by using the metric property of the measure, and theorem 17, that

$$\begin{aligned}\mu(A \cup B) &\geqslant \sup_n \mu(A \cup B_n) \\ &= \sup_n \{\mu(A) + \mu(B_n)\} \\ &= \mu(A) + \sup_n \mu(B_n) \\ &= \mu(A) + \mu(B),\end{aligned}$$

and we shall have the required measurability condition for F.

Define B_n to be the set of all points x of Ω in B with

$$\inf_{y \in F} \rho(x, y) > 1/n.$$

Then
$$B_1 \subset B_2 \subset B_3 \subset \ldots \subset B.$$

Further, if $b \in B$, then $b \notin F$, as $B \subset \Omega \backslash F$, and b is not a limit point of F and so lies in B_n for all sufficiently large n. Hence

$$B = \bigcup_{n=1}^{\infty} B_n.$$

As $A \subset F$, it follows, from the definition of B_n, that A and B_n are positively separated. Now consider any two points b and e with

$$b \in B_n, \qquad e \in \Omega \backslash B_{n+1}, \tag{8}$$

for some $n \geq 1$. Then, from the definition of B_{n+1},

$$\inf_{y \in F} \rho(e, y) \leq 1/(n+1).$$

So, for some choice of a point f in F, we have

$$\rho(e, f) \leq 1/(n + \tfrac{1}{2}).$$

If we had
$$\rho(b, e) \leq \frac{\frac{1}{2}}{n(n+\frac{1}{2})},$$

this would yield
$$\begin{aligned} \inf_{y \in F} \rho(b, y) &\leq \rho(b, f) \\ &\leq \rho(b, e) + \rho(e, f) \\ &\leq \frac{\frac{1}{2}}{n(n+\frac{1}{2})} + \frac{1}{n+\frac{1}{2}} \\ &= 1/n, \end{aligned}$$

contrary to the definition of B_n and the relation $b \in B_n$. Thus for all pairs b, e satisfying (8) we have $\rho(b, e) \geq 1/[n(2n+1)]$ and B_n and $\Omega \backslash B_{n+1}$ are positively separated. This completes the construction and so the whole proof.

***Theorem* 19.** *If μ is a metric measure in a metric space Ω all Borel sets in Ω are μ-measurable.*

Proof. Let $\mathscr{M}$ be the system of μ-measurable sets. Then $\mathscr{M}$ is a σ-field by theorem 3, and $\mathscr{M}$ contains the closed sets of Ω by theorem 18. Hence $\mathscr{M}$ contains the Borel sets by definition 8.

4.4. We now return to the study of the measures constructed by Method II and investigate their regularity and approximation properties.

***Theorem* 20.** *Let μ be the measure on a metric space Ω constructed, by Method II, from a pre-measure τ defined on a class $\mathscr{C}$ of sets with $\Omega \in \mathscr{C}$. Then μ is $\mathscr{C}_{\sigma\delta}$-regular.*

Proof. Let μ_δ be the measure on Ω constructed, by Method I, from the restriction τ_δ of τ to the sets C of $\mathscr{C}$ with $d(C) \leqslant \delta$. Then

$$\mu(E) = \sup_{\delta > 0} \mu_\delta(E).$$

By theorem 13, each measure μ_δ is $\mathscr{C}_{\sigma\delta}$-regular. So, if E is any set of Ω, for each positive integer n we can choose a $\mathscr{C}_{\sigma\delta}$-set C_n with

$$E \subset C_n, \qquad \mu_{1/n}(E) = \mu_{1/n}(C_n).$$

Then
$$C = \bigcap_{n \geqslant 1} C_n$$

is a $\mathscr{C}_{\sigma\delta}$-set with $E \subset C$, and, for each $\delta > 0$, there is an integer n with $0 < 1/n < \delta$ and

$$\mu_\delta(C) \leqslant \mu_{1/n}(C) \leqslant \mu_{1/n}(C_n) = \mu_{1/n}(E) \leqslant \mu(E).$$

Hence
$$\mu(C) \leqslant \mu(E),$$
so that we have

$$C \in \mathscr{C}_{\sigma\delta}, \qquad E \subset C, \qquad \mu(E) = \mu(C).$$

Thus μ is $C_{\sigma\delta}$-regular.

Corollary. *If τ is defined only on open sets, then μ is $\mathscr{G}_\delta$-regular. If τ is defined only on Borel sets, then μ is Borel-regular. In either case μ is regular.*

Proof. The first two statements follow by the (very simple) argument used to prove the corollary to theorem 13. The last statement follows by use of theorems 16 and 19.

4.5. The results on the regularity of measures constructed by Method II lead naturally to corresponding results on approximation from within.

***Theorem* 21.** *Let μ be the measure on a metric space Ω constructed, by Method II, from a pre-measure τ defined on a class $\mathscr{C}$ of sets with $\Omega \in \mathscr{C}$.*

Let E be a μ-measurable set with finite μ-measure. Then there is a set $\mathscr{C}$, of the form $C_1 \backslash C_2$ with C_1 and C_2 in $\mathscr{C}_{\sigma\delta}$, with

$$C \subset E, \qquad \mu(C) = \mu(E).$$

Proof. By theorem 20, the measure μ is $\mathscr{C}_{\sigma\delta}$-regular. The result follows by theorem 14.

Corollary. *If τ is defined only on open sets, then C can be taken to be of the form $C_1 \backslash C_2$ with C_1, C_2 both $\mathscr{G}_\delta$-sets. If τ is defined only on Borel sets, then C can be taken to be a Borel set.*

Proof. The results follow from the corollary to theorem 20 and theorem 14.

4.6. Our next aim is to prove a result, due to A.S. Besicovitch and P.A.P. Moran (1954) [see their lemma 3], that shows that C can be taken to be a $\mathscr{F}_\sigma$-set in the case when τ is defined only on open sets. For this we need a

Lemma. *In a metric space each open set is an $\mathscr{F}_\sigma$-set and each closed set is a $\mathscr{G}_\delta$-set.*

Proof. Let F be a closed set in a metric space Ω with metric ρ. Following the method used in the proof of theorem 18, we define G_n for $n = 1, 2, \ldots$ to be the set of points g, with

$$\inf_{y \in F} \rho(g, y) < 1/n.$$

Then, $F \subset G_n$ for each n. Also G_n is open for each n. Finally, if $e \notin F$ then

$$\inf_{y \in F} \rho(e, y)$$

is positive, as F is closed, and so e belongs to $\Omega \backslash G_n$ for all sufficiently large n. Hence

$$F = \bigcap_{n=1}^{\infty} G_n$$

and the closed set F is a $\mathscr{G}_\delta$-set. As the complement of a $\mathscr{G}_\delta$-set is automatically an $\mathscr{F}_\sigma$-set, it follows that each open set is an $\mathscr{F}_\sigma$-set.

Theorem 22. *Let μ be a $\mathscr{G}_\delta$-regular metric measure on a metric space Ω. If E is a μ-measurable set with finite μ-measure, there is an $\mathscr{F}_\sigma$-set C with*

$$C \subset E, \qquad \mu(C) = \mu(E),$$

and, if $\epsilon \geqslant 0$, there is a closed set F with

$$F \subset E, \qquad \mu(F) \geqslant \mu(E) - \epsilon.$$

Proof. The proof is a little complicated. Its plan is as follows. We approximate E from without by a $\mathscr{G}_\delta$-set

$$\bigcap_{n=1}^{\infty} G_n.$$

Each open set G_n is an $\mathscr{F}_\sigma$-set and so can be approximated by a closed set that covers 'most' of E. The intersection of these closed sets for $n \geqslant N$ will be a closed set B_N that approximates E in that

$$\mu(B_N \cap E) \geqslant \mu(E) - 2^{-N+1},$$

$$\mu(B_N \backslash E) \leqslant 2^{-N+1}.$$

Then $B_N \backslash E$ can be approximated from without by a $\mathscr{G}_\delta$-set and then, less accurately, but sufficiently accurately, by an open set $I_{NJ(N)}$. Thus $F_N = B_N \backslash I_{NJ(N)}$ will turn out to be a closed set approximating E, from within in that

$$F_N \subset E, \qquad \mu(F_N) > \mu(E) - 2^{-N+3},$$

and

$$C = \bigcup_{N=1}^{\infty} F_N$$

will be an $\mathscr{F}_\sigma$-set approximating E in the required way. In reading this proof, it is essential to remember that we are working with a measure, which may very well assign infinite measure to every non-empty open set. This difficulty is the source of many of the complications.

To carry out this programme let E be a μ-measurable set with finite μ-measure. As μ is $\mathscr{G}_\delta$-regular we can choose a $\mathscr{G}_\delta$-set G with

$$E \subset G, \qquad \mu(E) = \mu(G).$$

Then, for some sequence $\{G_n\}$ of open sets,

$$G = \bigcap_{n=1}^{\infty} G_n.$$

By replacing G_n by

$$\bigcap_{r=1}^{n} G_r,$$

if necessary, we may suppose that

$$G_1 \supset G_2 \supset G_3 \supset \ldots.$$

By the lemma we can write

$$G_n = \bigcup_{i=1}^{\infty} F_{ni} \quad (n = 1, 2, \ldots),$$

where the sets F_{ni}, $n, i = 1, 2, \ldots$, are all closed, and where we may suppose that

$$F_{n1} \subset F_{n2} \subset F_{n3} \subset \ldots \quad (n = 1, 2, \ldots).$$

As μ is a metric measure, the closed sets are μ-measurable (by theorem 18) and, for each fixed n, the sequence

$$E \cap F_{ni} \quad (i = 1, 2, \ldots)$$

is an increasing sequence of μ-measurable sets. By theorem 8,

$$\sup_i \mu(E \cap F_{ni}) = \mu\left(E \cap \bigcup_{i=1}^{\infty} F_{ni}\right) = \mu(E \cap G_n) = \mu(E) < +\infty.$$

Hence we can choose an integer $k(n)$ with

$$\mu(E \cap F_{nk(n)}) > \mu(E) - 2^{-n} \quad (n = 1, 2, \ldots).$$

As $F_{nk(n)}$ is μ-measurable, this gives

$$\mu(E) = \mu(E \cap F_{nk(n)}) + \mu(E \backslash F_{nk(n)}) > \mu(E) - 2^{-n} + \mu(E \backslash F_{nk(n)}),$$

so that

$$\mu(E \backslash F_{nk(n)}) < 2^{-n} \quad (n = 1, 2, \ldots). \tag{9}$$

We now take

$$B_N = \bigcap_{n \geqslant N} F_{nk(n)} \quad (N = 1, 2, \ldots)$$

as approximations, from within to

$$G = \bigcap_{n=1}^{\infty} \bigcup_{i=1}^{\infty} F_{ni}.$$

Clearly B_N is closed and

$$B_N \subset \bigcap_{n \geqslant N} G_n = G,$$

for all positive integers N. So

$$\mu(B_N) \leqslant \mu(G) = \mu(E) \quad (N = 1, 2, \ldots). \tag{10}$$

But, on the other hand,

$$\begin{aligned} E \cap B_N &= E \cap \left\{ \bigcap_{n \geqslant N} F_{nk(n)} \right\} \\ &= E \backslash \left\{ \bigcup_{n \geqslant N} (E \backslash F_{nk(n)}) \right\} \end{aligned}$$

so that, the sets all being μ-measurable with finite measure,

$$\begin{aligned} \mu(E \cap B_N) &= \mu(E) - \mu(\bigcup_{n \geqslant N} \{E \backslash F_{nk(n)}\}) \\ &\geqslant \mu(E) - \sum_{n \geqslant N} \mu(E \backslash F_{nk(n)}) \\ &\geqslant \mu(E) - \sum_{n \geqslant N} 2^{-n} \\ &= \mu(E) - 2^{-N+1}, \end{aligned}$$

on using (9). Using (10), we have

$$\mu(B_N \backslash E) + \mu(B_N \cap E) = \mu(B_N) \leqslant \mu(E) \leqslant \mu(E \cap B_N) + 2^{-N+1},$$

so that

$$\mu(B_N \backslash E) \leqslant 2^{-N+1}.$$

Hence B_N is a good approximation to E in that

$$\mu(B_N \cap E) \geqslant \mu(E) - 2^{-N+1}, \quad \mu(B_N \backslash E) \leqslant 2^{-N+1} \quad (N = 1, 2, \ldots). \tag{11}$$

As μ is $\mathscr{G}_\delta$-regular, we can choose a $\mathscr{G}_\delta$-set H_N with

$$B_N \backslash E \subset H_N, \quad \mu(B_N \backslash E) = \mu(H_N) \quad (N = 1, 2, \ldots).$$

We may suppose that

$$H_N = \bigcap_{j=1}^{\infty} I_{Nj}$$

with the sets I_{Nj} all open and

$$I_{N1} \supset I_{N2} \supset I_{N3} \supset \ldots \quad (N = 1, 2, \ldots).$$

Then, for each N, the sequence

$$E \cap I_{Nj} \quad (j = 1, 2, \ldots)$$

is a decreasing sequence of μ-measurable sets (as open sets are measurable for the metric measure μ by theorem 19) of finite μ-measure (as E has finite measure). Hence, by theorem 8,

$$\begin{aligned} \mu(E \cap H_N) &= \mu\left(E \cap \bigcap_{j=1}^{\infty} I_{Nj}\right) \\ &= \inf_j \mu(E \cap I_{Nj}) \quad (N = 1, 2, \ldots). \end{aligned}$$

So we can choose an integer $J(N)$ so that

$$\mu(E \cap I_{NJ(N)}) \leqslant \mu(E \cap H_N) + 2^{-N+1} \quad (N = 1, 2, \ldots).$$

Then, by (11),

$$\begin{aligned} \mu(E \cap I_{NJ(N)}) &\leqslant \mu(E \cap H_N) + 2^{-N+1} \\ &\leqslant \mu(H_N) + 2^{-N+1} \\ &= \mu(B_N \backslash E) + 2^{-N+1} \\ &\leqslant 2^{-N+2} \quad (N = 1, 2, \ldots). \end{aligned} \tag{12}$$

Now write

$$F_N = B_N \backslash I_{NJ(N)} \quad (N = 1, 2, \ldots).$$

Then F_N is closed,

$$\begin{aligned} F_N &= B_N \backslash I_{NJ(N)} \\ &\subset B_N \backslash H_N \\ &\subset B_N \backslash \{B_N \backslash E\} \\ &\subset E \quad (N = 1, 2, \ldots). \end{aligned}$$

Further, as $B_N \backslash E \subset I_{N(J(N))}$,

$$F_N = B_N \backslash I_{N\,J(N)}$$
$$\supset (E \cap B_N) \backslash (E \cap I_{N\,J(N)}),$$

and

$$\mu(F_N) \geqslant \mu((E \cap B_N) \backslash (E \cap I_{N\,J(N)}))$$
$$\geqslant \mu(E \cap B_N) - \mu(E \cap I_{N\,J(N)})$$
$$\geqslant \mu(E) - 2^{-N+1} - 2^{-N+2}$$
$$> \mu(E) - 2^{-N+3},$$

on using (11) and (12). So, if ϵ is positive, by taking N sufficiently large, we obtain a closed set F_N with

$$F_N \subset E, \qquad \mu(F_N) \geqslant \mu(E) - \epsilon.$$

Further the set

$$C = \bigcup_{N=1}^{\infty} F_N$$

is clearly an $\mathscr{F}_\sigma$-set with

$$C \subset E, \qquad \mu(C) = \mu(E).$$

This completes the proof.

Corollary. *Let μ be a measure on Ω. Let A_{ij} $(i, j = 1, 2, \ldots)$ be a system of μ-measurable sets. Suppose that the sets*

$$A_i = \bigcup_{j=1}^{\infty} A_{ij} \quad (i = 1, 2, \ldots)$$

form a decreasing sequence and that

$$A = \bigcap_{i=1}^{\infty} \bigcup_{j=1}^{\infty} A_{ij}$$

has finite μ-measure. Then there is an increasing sequence $k(1), k(2), \ldots$ of positive integers satisfying

$$\mu\Big(\bigcap_{i \geqslant N} \bigcup_{j \leqslant k(i)} A_{ij}\Big) \geqslant \mu(A) - 2^{-N+1},$$

for $N = 1, 2, \ldots$.

Proof. Write $E = A, G_i = A_i$, and $F_{ik} = \bigcup_{j=1}^{j=k} A_{ij}$. It suffices to repeat the argument leading to the inequality (11), noting that the various steps depend on the μ-measurability of the sets used rather than on their topological nature.

Remark. This corollary is essentially equivalent to Egoroff's Theorem; see Egoroff (1911) or M. E. Munroe (1953), p. 157, or S. Saks (1937), p. 17.

4.7. We can now summarise in one theorem most of the results that will be of importance in the succeeding chapters.

***Theorem* 23.** *Let τ be a pre-measure defined on a class $\mathscr{C}$ of sets in a metric space Ω. Suppose that each set of $\mathscr{C}$ is open. Let μ be the set function constructed from the pre-measure τ by Method II. Then μ is a regular, $\mathscr{G}_\delta$-regular metric measure, all Borel sets are μ-measurable, and each μ-measurable set of finite μ-measure contains an $\mathscr{F}_\sigma$-set with the same measure.*

Proof. By theorem 15, μ is a measure. By theorem 16, μ is a metric measure. By theorem 19, all Borel sets are μ-measurable. By the Corollary to theorem 20, μ is $\mathscr{G}_\delta$-regular. As Borel sets are μ-measurable, it follows that μ is regular. By theorem 22, each μ-measurable set of finite μ-measure contains an $\mathscr{F}_\sigma$-set with the same measure.

§5 Lebesgue measure in n-dimensional Euclidean space

In this section we relate the general theory we have developed to the special but most important Lebesgue measure.

Provided
$$a_i \leqslant b_i \quad (i = 1, 2, \ldots, n),$$
the set of points $\mathbf{x} = (x_1, x_2, \ldots, x_n)$
with
$$a_i < x_i < b_i \quad (i = 1, 2, \ldots, n),$$
will be called the open rectangle with corners
$$\mathbf{a} = (a_1, a_2, \ldots, a_n), \qquad \mathbf{b} = (b_1, b_2, \ldots, b_n),$$
and will be denoted by $R(\mathbf{a}, \mathbf{b})$. Let $\mathscr{R}$ be the class of all such rectangles and define a function τ on $\mathscr{R}$ by taking
$$\tau(R(\mathbf{a}, \mathbf{b})) = \prod_{i=1}^{n} (b_i - a_i).$$
Thus $\tau(R(\mathbf{a}, \mathbf{b}))$ is the elementary volume of the 'rectangle' $R(\mathbf{a}, \mathbf{b})$. Clearly τ is a pre-measure. Let λ, ν be the measures constructed from the pre-measure τ by Methods I and II. Then λ is easily seen to be the usual Lebesgue measure. But our theory suggests that the measure ν should have many advantages over the measure λ. But, in fact, λ shares all the advantages of ν as the first theorem shows.

***Theorem* 24.** *In the notation of this section $\lambda = \nu$.*

Proof. It is clear that for all sets E we have
$$\lambda(E) \leqslant \nu(E).$$

Suppose E is any set. To prove that

$$\nu(E) \leqslant \lambda(E)$$

it suffices to consider the case when $\lambda(E)$ is finite. If $\epsilon > 0$ is given, we can choose a sequence $\{R_i\}$ of rectangles of $\mathscr{R}$ with

$$E \subset \bigcup_{i=1}^{\infty} R_i, \qquad \sum_{i=1}^{\infty} \tau(R_i) < \lambda(E)+\epsilon.$$

Now, if $\delta > 0$, and $R = R(\mathbf{a}, \mathbf{b})$ is any rectangle, we can choose a large integer N, so that the N^n closed rectangles

$$a_i + \frac{r_i - 1}{N}(b_i - a_i) \leqslant x_i \leqslant a_i + \frac{r_i}{N}(b_i - a_i) \quad (i = 1, 2, \ldots, n),$$

obtained as $r_1, r_2, \ldots, r_n$ take all integral values with

$$1 \leqslant r_i \leqslant N \quad (i = 1, 2, \ldots, N),$$

all have diameter less than δ. Then provided η is sufficiently small and positive, the N^n rectangles $R(r_1, r_2, \ldots, r_n)$ given by

$$a_i + \frac{r_i - 1}{N}(b_i - a_i) - \eta < x_i < a_i + \frac{r_i}{N}(b_i - a_i) + \eta \quad (i = 1, 2, \ldots, n)$$

for $1 \leqslant r_i \leqslant N$ $(i = 1, 2, \ldots, N)$, cover $R(\mathbf{a}, \mathbf{b})$, have diameter less than δ and have

$$\sum_{r_1, r_2, \ldots, r_n = 1}^{N} \tau(R(r_1, r_2, \ldots, r_n)) \leqslant \tau(R(\mathbf{a}, \mathbf{b})) + O(\eta).$$

Applying this process to each rectangle R_i we can replace R_i by a finite collection, say R_{ij} $(i = 1, 2, \ldots, j(i))$, of rectangles of $\mathscr{R}$ of diameter less than δ, covering R_i and with

$$\sum_{j=1}^{j(i)} \tau(R_{ij}) \leqslant \tau(R_i) + \epsilon 2^{-i}.$$

Then the system of all such rectangles R_{ij} is a cover of E by sets of diameter less than δ with

$$\Sigma\tau(R_{ij}) \leqslant \lambda(E) + 2\epsilon.$$

Hence
$$\inf_{\substack{\cup R \subset E \\ d(R) \leqslant \delta}} \Sigma\tau(R_i) \leqslant \lambda(E) + 2\epsilon,$$

so
$$\nu(E) \leqslant \lambda(E) + 2\epsilon.$$

As ϵ may be any positive number we have $\nu(E) \leqslant \lambda(E)$ as required.

It is now clear that theorem 23 applies and gives some non-trivial properties of Lebesgue measure. But there is one basic property of

Lebesgue measure that escapes the scope of the general theory. We prove a result, originally proved, in the one-dimensional case, by Borel (1895).

***Theorem* 25.** *In the notation of this section, for any rectangle R of $\mathscr{R}$,*

$$\lambda(R) = \tau(R).$$

Proof. It is clear that for any open rectangle R we have

$$\lambda(R) \leqslant \tau(R).$$

We first consider a closed rectangle I of the form

$$a_i \leqslant x_i \leqslant b_i \quad (i = 1, 2, \ldots, n),$$

where
$$a_i < b_i \quad (i = 1, 2, \ldots, n).$$

Let $\{R_j\}$ be any sequence of rectangles of $\mathscr{R}$ with

$$I \subset \bigcup_{j=1}^{\infty} R_j.$$

Then, as I is closed and bounded, and as each R_j is open, there is an integer N with

$$I \subset \bigcup_{j=1}^{N} R_j.$$

Let R be the interior of I and let $R \cap R_j$ be the rectangle

$$a_i^{(j)} < x_i < b_i^{(j)} \quad (i = 1, 2, \ldots, n),$$

for $j = 1, 2, \ldots, N$. For $i = 1, 2, \ldots, n$, let

$$a_i = c_i^{(1)}, c_i^{(2)}, \ldots, c_i^{(2N)} = b_i$$

be a rearrangement of the numbers

$$a_i^{(1)}, b_i^{(1)}, \ldots, a_i^{(N)}, b_i^{(N)}$$

in non-decreasing order.

For any sequence $k(1), k(2), \ldots, k(n)$ with $0 \leqslant k(i) \leqslant 2N-1$, consider the rectangle $R(k(1), k(2), \ldots, k(n))$ of points $(x_1, x_2, \ldots, x_n)$ with

$$c_i^{(k(i))} < x_i < c_i^{(k(i)+1)} \quad (i = 1, 2, \ldots, n).$$

When this rectangle is non-empty its mid-point is in R and so is in one of the rectangles R_j, $1 \leqslant j \leqslant N$. Hence, by the choice of the numbers $c_i^{(k)}$, the whole rectangle $R(k(1), k(2), \ldots, k(n))$ lies in one of the rectangles $R \cap R_j$. But, by the choice of numbers $c_i^{(k)}$, the elementary volume of any rectangle R_j is the sum of the elementary volumes of the rectangles $R(k(1), k(2), \ldots, k(n))$ that it contains. Hence

$$\begin{aligned}\sum_{j=1}^{\infty} \tau(R_j) &\geqslant \sum_{j=1}^{N} \tau(R_j) \\ &\geqslant \sum_{j=1}^{N} \tau(R \cap R_j) \\ &= \sum_{j=1}^{N} \sum_{R(k(1),\ldots,k(n)) \subset R \cap R_j} \prod_{i=1}^{n} (c_i^{(k(i)+1)} - c_i^{(k(i))}) \\ &\geqslant \sum_{k(1),\ldots,k(n)=1}^{2N-1} \prod_{i=1}^{n} (c_i^{(k(i)+1)} - c_i^{(k(i))}) \\ &= \prod_{i=1}^{n} \{c_i^{(2N)} - c_i^{(1)}\} = \prod_{i=1}^{n} (b_i - a_i).\end{aligned}$$

Hence
$$\lambda(I) \geqslant \prod_{i=1}^{n} (b_i - a_i).$$

Now given any non-empty open rectangle R we can construct in R closed rectangles with elementary volume, and hence with Lebesgue measure, as close as we please to $\tau(R)$. So we must have $\lambda(R) \geqslant \tau(R)$. Hence $\lambda(R) = \tau(R)$ as required.

§6 Metric measures in topological spaces

The classical definition that we have used for the concept of a metric measure depends essentially on the metric of the space. It would be desirable to have a definition of a 'metric' measure which was expressed in purely topological terms (rather than in terms of a metric) which would have a natural meaning in a topological space and which would yield a class of measures in topological spaces which would have properties similar to the metric measures in metric spaces. In particular, one would hope that the Borel sets would turn out to be measurable for such measures, and that in appropriate circumstances this would lead to regularity results. It is certainly possible to develop such a theory. Indeed there are perhaps too many theories of this type. They have been developed by W. W. Bledsoe and A. P. Morse (1963), by C. A. Rogers and M. Sion (1963), by J. D. Knowles (1966), by M. Sion and R. C. Willmott (1966), and by R. C. Willmott (1967). As the theories lie at a level of abstraction out of keeping with the rest of this book we carry the matter no further.

§7 The Souslin operation

In this section, which should perhaps logically belong to §2 on measures in abstract spaces, but which has been postponed to this section as it introduces some quite new ideas, our object is to show that the class of sets measurable with respect to a given measure is closed under the Souslin operation.

7.1. If $\mathscr{A}$ is any class of sets, the minimal σ-field containing $\mathscr{A}$ is a somewhat inaccessible concept; this minimal σ-field can be 'constructed' either by taking the intersection of *all* the σ-fields containing $\mathscr{A}$ or alternatively by an uncountable transfinite extension process applied to $\mathscr{A}$. There is another class of sets associated with $\mathscr{A}$, the class of Souslin-$\mathscr{A}$ sets; a class that is in some ways simpler and in other ways more recondite.

***Definition* 15.** *If $\mathscr{A}$ is any class of sets, the Souslin-$\mathscr{A}$ sets are the sets of the form*

$$A = \bigcup_{i_1, i_2, \ldots} \bigcap_{n=1}^{\infty} A_{i_1, i_2, \ldots, i_n} \qquad (1)$$

with $A_{i_1, i_2, \ldots, i_n} \in \mathscr{A}$ for every infinite sequence $i_1, i_2, \ldots, i_n$ of positive integers the union being over all finite sequences of positive integers. The set A is said to result from application of the Souslin operation to the system of sets $A_{i_1, i_2, \ldots, i_n}$.

It must be noted that although the set A is built up from the countable system of sets $A_{i_1, i_2, \ldots, i_n}$ of $\mathscr{A}$, the union in (1) is over the continuum many infinite sequences of positive integers.

As the notation used in (1) is rather difficult to write, to print and to read we introduce some conventions. We use $\mathbf{I}$ to denote the system of all infinite vectors

$$\mathbf{i} = i_1, i_2, \ldots,$$

with positive integral elements. Given such an $\mathbf{i}$ in $\mathbf{I}$ we use $\mathbf{i}|n$, read '$\mathbf{i}$ restricted to n', to denote the finite vector

$$\mathbf{i}|n = i_1, i_2, \ldots, i_n$$

obtained by taking the first n components of $\mathbf{i}$.

In terms of these conventions, if $\mathscr{A}$ is any class of sets, the Souslin-$\mathscr{A}$ sets are the sets of the form

$$A = \bigcup_{\mathbf{i} \in \mathbf{I}} \bigcap_{n=1}^{\infty} A_{\mathbf{i}|n},$$

where $A_{\mathbf{i}|n} \in \mathscr{A}$ for all $n \geqslant 1$ and $\mathbf{i}$ in $\mathbf{I}$. Here, of course, the notation implies that $A_{\mathbf{i}|n}$, although it naturally depends on the first n components of $\mathbf{i}$, is independent of all the remaining components.

7.2. Our aim is to establish the result (that has apparently grown out of the original result of N. Lusin (1917) that analytic sets are Lebesgue measurable) that the class of sets measurable for a given measure is closed under the Souslin operation (see also S. Saks, 1937 and E. J. Mickle, 1958). We first prove a lemma showing that this result holds for regular measures.

Lemma. *Let μ be a regular measure and let $\mathcal{M}$ be the class of μ-measurable sets. Then all Souslin-$\mathcal{M}$ sets are μ-measurable.*

Proof. Let

$$E = \bigcup_{\mathbf{i} \in \mathbf{I}} \bigcap_{n=1}^{\infty} M_{\mathbf{i}|n},$$

where $M_{\mathbf{i}|n} \in \mathcal{M}$ for all $n \geq 1$ and for all $\mathbf{i} \in \mathbf{I}$. We first restandardize the representation of E by writing

$$N(\mathbf{i}|n) = \bigcap_{k=1}^{n} M_{\mathbf{i}|k},$$

for all $n \geq 1$ and all $\mathbf{i}$ in $\mathbf{I}$. Then

$$E = \bigcup_{\mathbf{i} \in \mathbf{I}} \bigcap_{n=1}^{\infty} N(\mathbf{i}|n), \tag{2}$$

where $N(\mathbf{i}|n) \in \mathcal{M}$ for all $n \geq 1$ and for all $\mathbf{i} \in \mathbf{I}$.

We suppose that A, B are two sets of finite μ-measure with

$$A \subset E, \qquad B \subset \Omega \backslash E.$$

By theorem 1, it will suffice to prove that

$$\mu(A \cup B) \geq \mu(A) + \mu(B).$$

Let $\epsilon > 0$ be given. Let $\mathbf{I}(1;k)$ denote the set of all $\mathbf{i} = i_1, i_2, \ldots$ of $\mathbf{I}$ with $1 \leq i_1 \leq k$. Then

$$E = \bigcup_{k=1}^{\infty} \bigcup_{\mathbf{i} \in \mathbf{I}(1;k)} \bigcap_{n=1}^{\infty} N(\mathbf{i}|n)$$

and the sequence

$$A \cap \bigcup_{\mathbf{i} \in \mathbf{I}(1;k)} \bigcap_{n=1}^{\infty} N(\mathbf{i}|n) \quad (k = 1, 2, \ldots)$$

is a non-decreasing sequence of sets with union $A \cap E = A$. As μ is a regular measure, it follows from theorem 9, that we can choose a positive integer k_1 with

$$\mu(A \cap \bigcup_{\mathbf{i} \in \mathbf{I}(1;k)} \bigcap_{n=1}^{\infty} N(\mathbf{i}|n)) > \mu(A) - \epsilon,$$

for $k \geq k_1$. Write $\mathbf{K}(1) = \mathbf{I}(1; k_1)$. Then

$$\mu(A \cap \bigcup_{\mathbf{i} \in \mathbf{K}(1)} \bigcap_{n=1}^{\infty} N(\mathbf{i}|n)) > \mu(A) - \epsilon,$$

and $\mathbf{K}(1)$ is the set of $\mathbf{i}$ in $\mathbf{I}$ with $1 \leq i_1 \leq k_1$.

We now suppose that for some $r \geq 1$, positive integers $k_1, k_2, \ldots, k_r$ have been chosen so that

$$\mu\left(A \cap \bigcup_{\mathbf{i} \in \mathbf{K}(r)} \bigcap_{n=1}^{\infty} N(\mathbf{i}|n)\right) > \mu(A) - \epsilon, \tag{3}$$

where $\mathbf{K}(r)$ is the set of $\mathbf{i}$ in $\mathbf{I}$ with

$$1 \leq i_1 \leq k_1, \quad 1 \leq i_2 \leq k_2, \ldots, \quad 1 \leq i_r \leq k_r.$$

Let $\mathbf{I}(r+1; k)$ denote the set of $\mathbf{i}$ in $\mathbf{I}$ with

$$1 \leq i_1 \leq k_1, 1 \leq i_2 \leq k_2, \ldots, 1 \leq i_r \leq k_r, 1 \leq i_{r+1} \leq k.$$

Then the sequence

$$A \cap \bigcup_{\mathbf{i} \in \mathbf{I}(r+1; k)} \bigcap_{n=1}^{\infty} N(\mathbf{i}|n) \quad (k = 1, 2, \ldots)$$

is a non-decreasing sequence of sets with union

$$A \cap \bigcup_{\mathbf{i} \in \mathbf{K}(r)} \bigcap_{n=1}^{\infty} N(\mathbf{i}|n),$$

so we can choose k_{r+1} large enough to ensure that

$$\mu\left(A \cap \bigcup_{\mathbf{i} \in \mathbf{K}(r+1)} \bigcap_{n=1}^{\infty} N(\mathbf{i}|n)\right) > \mu(A) - \epsilon,$$

where $\mathbf{K}(r+1)$ is the set of $\mathbf{i}$ in $\mathbf{I}$ with $1 \leq i_1 \leq k_1$, $1 \leq i_2 \leq k_2, \ldots,$ $1 \leq i_{r+1} \leq k_{r+1}$. So we may suppose that $k_1, k_2, \ldots$ have been chosen inductively in this way to ensure that (3) holds for $r = 1, 2, \ldots$.

We now transfer our attention to another system of sets. For each $r > 1$ we have

$$\bigcup_{\mathbf{i} \in \mathbf{K}(r)} \bigcap_{n=1}^{\infty} N(\mathbf{i}|n) \subset \bigcup_{\mathbf{i} \in \mathbf{K}(r)} N(\mathbf{i}|r)$$

so that

$$\mu(A) - \epsilon < \mu\left(A \cap \bigcup_{\mathbf{i} \in \mathbf{K}(r)} \bigcap_{n=1}^{\infty} N(\mathbf{i}|n)\right) \leq \mu\left(A \cap \bigcup_{\mathbf{i} \in \mathbf{K}(r)} N(\mathbf{i}|r)\right).$$

But, as $\mathbf{i}$ runs over $\mathbf{K}(r)$, the r-vector $\mathbf{i}|r$ takes only a countable number of different values. So, for each r,

$$\bigcup_{\mathbf{i} \in \mathbf{K}(r)} N(\mathbf{i}|r)$$

is μ-measurable. Further, as

$$\mathbf{K}(r+1) \subset \mathbf{K}(r), \qquad N(\mathbf{i}|r+1) \subset N(\mathbf{i}|r)$$

for all $\mathbf{i}$ in $\mathbf{I}$ and $r \geqslant 1$, the sequence

$$\bigcup_{\mathbf{i} \in \mathbf{K}(r)} N(\mathbf{i}|r) \quad (r = 1, 2, \ldots),$$

is a non-increasing sequence of μ-measurable sets. By theorem 11, these are also measurable for the measure μ_A defined by the formula

$$\mu_A(S) = \mu(A \cap S),$$

for all S of Ω. As μ_A takes only finite values, it follows from theorem 8, that

$$\begin{aligned} \mu_A\left(\bigcap_{r=1}^{\infty} \bigcup_{\mathbf{i} \in \mathbf{K}(r)} N(\mathbf{i}|r)\right) &= \inf_r \mu_A\left(\bigcup_{\mathbf{i} \in \mathbf{K}(r)} N(\mathbf{i}|r)\right) \\ &= \inf_r \mu\left(A \cap \bigcup_{\mathbf{i} \in \mathbf{K}(r)} N(\mathbf{i}|r)\right) \\ &\geqslant \mu(A) - \epsilon. \end{aligned}$$

Thus, writing

$$N = \bigcap_{r=1}^{\infty} \bigcup_{\mathbf{i} \in \mathbf{K}(r)} N(\mathbf{i}|r),$$

we have a μ-measurable set N with

$$\mu(A \cap N) \geqslant \mu(A) - \epsilon.$$

We now use a set-theoretic result:

$$N \subset E; \tag{4}$$

to whose proof we will return after we have completed the measure-theoretic part of the proof. Using (4) we have

$$\{A \cap N\} \subset N, \quad B \subset \Omega \backslash N,$$

so that, by the μ-measurability of N,

$$\begin{aligned} \mu(A \cup B) &\geqslant \mu(\{A \cap N\} \cup B) \\ &= \mu(A \cap N) + \mu(B) \\ &\geqslant \mu(A) - \epsilon + \mu(B). \end{aligned}$$

As ϵ was any positive number this implies

$$\mu(A \cup B) \geqslant \mu(A) + \mu(B),$$

and the proof will be complete when we have established (4).

Consider any point x of

$$N = \bigcap_{r=1}^{\infty} \bigcup_{\mathbf{i} \in \mathbf{K}(r)} N(\mathbf{i}|r).$$

Then, for each $r \geqslant 1$, we can choose a vector $\mathbf{i}(r)$ of $\mathbf{K}(r)$ with

$$x \in N(\mathbf{i}(r)|r).$$

As r takes the values $1, 2, \ldots$, the first component $i_1(r)$ of $\mathbf{i}(r)$ takes only the values i_1 with $1 \leqslant i_1 \leqslant k_1$, and so we can choose an integer j_1 with $1 \leqslant j_1 \leqslant k_1$ and an infinite sequence R_1 so that

$$i_1(r) = j_1 \quad \text{for} \quad r \in R_1.$$

Proceeding inductively in this way we can choose integers $j_1, j_2, \ldots$, and infinite subsequences $R_1, R_2, \ldots$ so that

$$1 \leqslant j_s \leqslant k_s \quad (s = 1, 2, \ldots),$$

$$R_1 \supset R_2 \supset R_3 \supset \ldots,$$

$$i_s(r) = j_s \quad \text{for} \quad r \in R_s \quad (s = 1, 2, \ldots).$$

Write

$$\mathbf{j} = j_1, j_2, \ldots.$$

Now, for each integer s, there is an integer $r \geqslant s$ in R_s. Then

$$i_t(r) = j_t \quad \text{for} \quad 1 \leqslant t \leqslant s.$$

So

$$x \in N(\mathbf{i}(r)|r) \subset N(\mathbf{i}(r)|s) = N(\mathbf{j}|s).$$

Thus

$$x \in \bigcap_{s=1}^{\infty} N(\mathbf{j}|s) \subset \bigcup_{\mathbf{i} \in \mathbf{I}} \bigcap_{n=1}^{\infty} N(\mathbf{i}|n) = E.$$

Consequently $N \subset E$ as required.

***Theorem* 26.** *Let μ be any measure and let $\mathscr{M}$ be the class of all μ-measurable sets. Then all Souslin-$\mathscr{M}$ sets are μ-measurable.*

Proof. Let S be a Souslin-$\mathscr{M}$ set. Consider any two sets A, B of finite μ-measure with

$$A \subset S, \qquad B \subset \Omega \backslash S.$$

We first introduce the set function ν defined by

$$\nu(E) = \mu((A \cup B) \cap E).$$

By theorem 11 this set function is a measure and all μ-measurable sets are ν-measurable. Further

$$\nu(\Omega \backslash (A \cup B)) = \mu(\varnothing) = 0,$$

so that $\Omega \backslash (A \cup B)$ and $A \cup B$ are ν-measurable. Now let λ be the measure constructed by Method I from the restriction of ν to its class of ν-measurable sets. By theorem 7, λ is a regular measure and all ν-measurable sets are λ-measurable. As $A \cup B$ is ν-measurable we have

$$\lambda(A \cup B) = \nu(A \cup B) = \mu(A \cup B), \tag{5}$$

also

$$\lambda(A) \geqslant \nu(A) = \mu(A), \tag{6}$$

$$\lambda(B) \geqslant \nu(B) = \mu(B). \tag{7}$$

As all sets of $\mathscr{M}$ are λ-measurable, and λ is a regular measure, it follows by the lemma that S is λ-measurable. Hence

$$\lambda(A \cup B) = \lambda(A) + \lambda(B).$$

Using this with (5), (6) and (7) we obtain

$$\mu(A \cup B) \geqslant \mu(A) + \mu(B).$$

This completes the proof.

2

HAUSDORFF MEASURES

§1 Definition of Hausdorff measures and equivalent definitions

We start with a formal definition of a Hausdorff measure as a certain very special type of Method II measure. We note that the results of Chapter 1 give quite a lot of information about these measures. We show that the same measures can be obtained by a number of different processes.

1.1. Throughout the chapter Ω will denote a metric space with metric ρ. We use $\mathscr{H}$ to denote the class of functions h defined for all $t \geqslant 0$, but perhaps having the value $+\infty$ for some values of t, monotonic increasing for $t \geqslant 0$, positive for $t > 0$ and continuous on the right for all $t \geqslant 0$. We will use $\mathscr{H}_0$ for the subset of all h of $\mathscr{H}$ with $h(0) = 0$.

***Definition* 16.** *Let Ω be a metric space and let h be a function of $\mathscr{H}$. Define $h(G)$ for G an open set of Ω by $h(G) = h(d(G))$ if $G \neq \varnothing$ and $h(\varnothing) = 0$. Then the measure constructed from the pre-measure h, defined on $\mathscr{G}$, by Method II is called the Hausdorff measure corresponding to the function h, or simply h-measure, and is denoted by μ^h.*

Note that it is easy to verify that h is indeed a pre-measure, so that the definition is valid. One special case is when $h(t) = 1$ for all $t \geqslant 0$, it is an easy exercise to verify that in this case $\mu^h(E)$ is equal to the number of points of E, if this number is finite, and $+\infty$, if the number of points of E is infinite.

Specializing theorem 23 we have immediately

***Theorem* 27.** *A Hausdorff measure μ^h is a regular, $\mathscr{G}_\delta$-regular metric measure, all Borel sets are μ^h-measurable, and each μ^h-measurable set of finite μ^h-measure contains an $\mathscr{F}_\sigma$-set with the same measure.*

1.2. In order to obtain theorem 27 directly from theorem 23, it is clearly convenient to define $\mu^h(E)$ in terms of open covers of E. In some arguments it is convenient to work with these open covers, in other arguments it is convenient to work with closed covers or with covers that are subsets of E. Our next theorem shows that we can obtain the same measure by any one of these methods. We now use $h(E)$ quite generally as an abbreviation for $h(d(E))$, when $E \neq \varnothing$, and for 0 when $E = \varnothing$.

***Theorem* 28.** *Let $h \in \mathscr{H}$ and let E be a set of Ω. For each $\delta > 0$, let*

$$\mu_\delta^h(E) = \inf_{\substack{\cup G \supset E \\ G \in \mathscr{G} \\ d(G) \leqslant \delta}} \Sigma h(G_i),$$

$$\nu_\delta^h(E) = \inf_{\substack{\cup F \supset E \\ F \in \mathscr{F} \\ d(F) \leqslant \delta}} \Sigma h(F_i),$$

$$\sigma_\delta^h(E) = \inf_{\substack{\cup S \supset E \\ d(S) \leqslant \delta}} \Sigma h(S_i),$$

$$\tau_\delta^h(E) = \inf_{\substack{\cup S = E \\ d(S) \leqslant \delta}} \Sigma h(S_i),$$

the sets $\{S_i\}$ in the definition of σ_δ^h and τ_δ^h being arbitrary sets of Ω. Then, for any ϵ with $\epsilon > \delta$, we have

$$\mu_\epsilon^h(E) \leqslant \nu_\delta^h(E) = \sigma_\delta^h(E) = \tau_\delta^h(E) \leqslant \mu_\delta^h(E). \tag{1}$$

Further

$$\mu^h(E) = \sup_{\delta>0} \mu_\delta^h(E) = \sup_{\delta>0} \nu_\delta^h(E) = \sup_{\delta>0} \sigma_\delta^h(E) = \sup_{\delta>0} \tau_\delta^h(E). \tag{2}$$

Proof. The inequality $\mu_\epsilon^h(E) \leqslant \nu_\delta^h(E)$ is trivial if $\nu_\delta^h(E) = +\infty$. So we suppose that $\nu_\delta^h(E)$ is finite. Then, for each $\eta > 0$, we can choose a sequence $\{F_i\}$ of closed sets with

$$\cup F_i \supset E, \qquad d(F_i) \leqslant \delta, \qquad \Sigma h(F_i) \leqslant \nu_\delta^h(E) + \eta.$$

If $F_i = \varnothing$ we write $G_i = \varnothing$ and have $d(G_i) < \delta$, $h(G_i) = 0 = h(F_i)$. If $F_i \neq \varnothing$, we use the continuity of h on the right at $t = d(F_i)$ to choose $\eta_i > 0$ so small that

$$h(d(F_i) + 2\eta_i) < h(F_i) + \eta 2^{-i},$$

$$\delta + 2\eta_i < \epsilon.$$

Then taking G_i to be the set of all points at a distance strictly less than η_i from some point of F_i we have an open set G_i with

$$F_i \subset G_i,$$

$$d(G_i) \leqslant d(F_i) + 2\eta_i \leqslant \delta + 2\eta_i < \epsilon,$$

$$h(G_i) \leqslant h(d(F_i) + 2\eta_i) \leqslant h(F_i) + \eta 2^{-i}.$$

This construction yields a sequence $\{G_i\}$ of open sets with

$$\bigcup G_i \supset \bigcup F_i \supset E, \qquad d(G_i) \leqslant \epsilon,$$

$$\Sigma h(G_i) \leqslant \Sigma h(F_i) + \Sigma \eta 2^{-i} \leqslant \nu_\delta^h(E) + 2\eta.$$

As η was arbitrary, we have completed the proof that

$$\mu_\epsilon^h(E) \leqslant \nu_\delta^h(E).$$

Now the conditions

$$\bigcup S_i \supset E, \qquad S_i \in \mathscr{F}, \qquad d(S_i) \leqslant \delta,$$

are more restrictive on the sequence $\{S_i\}$ than the conditions

$$\bigcup S_i \supset E, \qquad d(S_i) \leqslant \delta.$$

Hence $$\sigma_\delta^h(E) = \inf_{\substack{\cup S \supset E \\ d(S) \leqslant \delta}} \Sigma h(S_i) \leqslant \inf_{\substack{\cup S \supset E \\ S \in \mathscr{F} \\ d(S) \leqslant \delta}} \Sigma h(S_i) = \nu_\delta^h(E).$$

But on the other hand, if S_i is any set, we can take F_i to be the closure of S_i and we will have $F_i = \varnothing$ if and only if $S_j = \varnothing$, and

$$d(F_i) = d(S_i)$$

so that

$$S_i \subset F_i, \qquad h(F_i) = h(S_i), \qquad d(F_i) = d(S_i).$$

Hence we obtain

$$\nu_\delta^h(E) = \inf_{\substack{\cup F \supset E \\ F \in \mathscr{F} \\ d(F) \leqslant \delta}} \Sigma h(F_i) \leqslant \inf_{\substack{\cup S \supset E \\ d(S) \leqslant \delta}} \Sigma h(S_i) = \sigma_\delta^h(E).$$

Consequently $$\nu_\delta^h(E) = \sigma_\delta^h(E),$$
as required.

Similarly we obtain $$\sigma_\delta^h(E) \leqslant \tau_\delta^h(E),$$

trivially and $$\tau_\delta^h(E) \leqslant \sigma_\delta^h(E),$$

on noting that, if $\{S_i\}$ is any sequence of sets with

$$E \subset \bigcup S_i, \qquad d(S_i) \leqslant \delta,$$

the sequence of sets $T_i = E \cap S_i$, $i = 1, 2, \ldots$, has the properties

$$E = \bigcup T_i, \qquad d(T_i) \leqslant \delta, \qquad \Sigma h(T_i) \leqslant \Sigma h(S_i).$$

Hence $\sigma_\delta^h(E) = \tau_\delta^h(E)$.

The inequality $$\sigma_\delta^h(E) \leqslant \mu_\delta^h(E)$$

is again trivial. So we have proved (1). The result (2) follows immediately from (1).

Remark. The formulae

$$\mu^h(E) = \sup_{\delta>0} \tau_\delta^h(E),$$

$$\tau_\delta^h(E) = \inf_{\substack{\cup S=E \\ d(S)\leqslant\delta}} \Sigma h(S_i),$$

show that the μ^h-measure can be defined in terms of the diameters and covering properties of the subsets of E. Thus the μ^h-measure of a set E is an intrinsic property of E as a set on which a metric function ρ is defined. Thus, if E is removed from Ω and re-embedded in a metric space Ω' with metric ρ' so that $\rho' = \rho$ on E, this will not change the μ^h-measure of E.

§2 Mappings, special Hausdorff measures, surface areas

The object of this section is to establish a connection between the Hausdorff measure corresponding to the function t^r, r a positive integer, and the theory of the 'area' of r-dimensional surfaces. We shall call this measure (r)-measure and denote it by $\mu^{(r)}$. After studying the effect of continuous mappings, satisfying Lipschitz conditions, on Hausdorff measures we establish the connection between (r)-measure in r-dimensional Euclidean space and Lebesgue measure in the same space. These results enable us to show that the image in n-dimensional Euclidean space, under a mapping that is smooth with a smooth inverse, of a bounded open set in r-dimensional Euclidean space has finite positive (r)-measure. This provides the link with the theory of surface areas.

2.1. We start with the mapping theorem.

Theorem 29. *Let f be a function defined on a set E of a metric space Ω' with metric ρ'. Let f take values in a metric space Ω with metric ρ. Let g be a continuous strictly increasing function defined for $t \geqslant 0$ and with $g(0) = 0$. Suppose that f satisfies the Lipschitz condition*

$$\rho(f(x), f(y)) \leqslant g(\rho'(x, y)),$$

for all x, y in E. Then, for all h in $\mathscr{H}$ and all $\delta > 0$,

$$\mu_{g(\delta)}^{hg^{-1}}(f(E)) \leqslant \mu_\delta^h(E), \qquad \mu^{hg^{-1}}(f(E)) \leqslant \mu^h(E).$$

Proof. First, note that as g is continuous and strictly increasing for $t \geqslant 0$ and $g(0) = 0$, the inverse function g^{-1} will be defined, perhaps with the value $+\infty$, continuous, strictly increasing with $g^{-1}(0) = 0$. Hence, for all h in $\mathscr{H}$, hg^{-1} is also in $\mathscr{H}$.

Now fix h in $\mathscr{H}$ and $\delta > 0$. If $\{S_i\}$ is any sequence of sets of Ω' with

$$\bigcup S_i \supset E, \qquad S_i \subset E. \qquad d'(S_i) \leqslant \delta,$$

for each i, using d' for diameters in Ω', then

$$\begin{aligned} d(f(S_i)) &= \sup_{x,\, y \in S_i} \rho(f(x), f(y)) \\ &\leqslant \sup_{x,\, y \in S_i} g(\rho'(x, y)) \\ &\leqslant g(d'(S_i)) \leqslant g(\delta), \end{aligned}$$

and
$$hg^{-1}\{d(f(S_i))\} \leqslant hg^{-1}g(d'(S_i)) = h(S_i).$$

It follows immediately, from theorem 28 that

$$\mu_{g(\delta)}^{hg^{-1}}(f(E)) \leqslant \mu_\delta^h(E).$$

Taking the supremum over all values of δ on each side of this inequality, we obtain

$$\mu^{hg^{-1}}(f(E)) \leqslant \mu^h(E).$$

2.2. We now establish a connection between (n)-measure in Euclidean space E_n and Lebesgue measure in E_n.

***Theorem* 30.** *For each natural number n there is a finite positive constant κ_n so that, for every set E in Euclidean space E_n we have*

$$\mu^{(n)}(E) = \kappa_n \lambda(E),$$

λ denoting the Lebesgue measure in E_n.

Proof. We first study the cube C_0 of all points $\mathbf{x} = (x_1, x_2, \ldots, x_n)$ with

$$0 \leqslant x_1 < 1, \qquad 0 \leqslant x_2 < 1, \ldots, \qquad 0 \leqslant x_n < 1.$$

If $\delta > 0$ is given, we can choose an integer N so large that

$$(\sqrt{n})/N < \delta.$$

Now C_0 is covered by the system of N^n cubes defined by the inequalities

$$\frac{r_1 - 1}{N} \leqslant x_1 < \frac{r_1}{N}, \quad \frac{r_2 - 1}{N} \leqslant x_2 < \frac{r_2}{N}, \ldots, \quad \frac{r_n - 1}{N} \leqslant x_n < \frac{r_n}{N},$$

as $r_1, r_2, \ldots, r_n$ take all the possible integral values from 1 to N. As these cubes all have the diameter $(\sqrt{n})/N$ which is less than δ, it follows that

$$\mu_\delta^{(n)}(C_0) \leqslant \sum_{r_1, r_2, \ldots, r_n = 1}^{N} \left(\frac{\sqrt{n}}{N}\right)^n = n^{\frac{1}{2}n}.$$

Hence
$$\mu^{(n)}(C_0) \leqslant n^{\frac{1}{2}n}.$$

Now suppose $\{S_i\}$ is any sequence of sets covering C_0. For each i we can choose a closed cube C_i containing S_i with the edge of C_i equal to twice the diameter of S_i. Then, using the properties of Lebesgue measure, we have

$$C_0 \subset \bigcup_{i=1}^{\infty} S_i \subset \bigcup_{i=1}^{\infty} C_i,$$

and
$$1 = \lambda(C_0) \leqslant \lambda\left(\bigcup_{i=1}^{\infty} C_i\right) \leqslant \sum_{i=1}^{\infty} \lambda(C_i) = \sum_{i=1}^{\infty} \{2d(S_i)\}^n,$$

so that
$$\sum_{i=1}^{\infty} \{d(S_i)\}^n \geqslant 2^{-n}.$$

As this inequality holds for each cover $\{S_i\}$ of C_0 we deduce that

$$\mu^{(n)}(C_0) \geqslant 2^{-n}.$$

Now we write
$$\kappa_n = \mu^{(n)}(C_0),$$

and have
$$2^{-n} \leqslant \kappa_n \leqslant n^{\frac{1}{2}n}.$$

Then
$$\mu^{(n)}(C_0) = \kappa_n \lambda(C_0).$$

By the invariance of $\mu^{(n)}$ and λ under translation and as each is homogeneous of degree n under similarity transformations, we deduce that

$$\mu^{(n)}(C) = \kappa_n \lambda(C),$$

for all cubes C defined by inequalities of the form

$$a_1 \leqslant x_1 < a_1 + s, \quad a_2 \leqslant x_2 < a_2 + s, \ldots, \quad a_n \leqslant x_n < a_n + s.$$

We now consider a special class $\mathscr{C}$ of all cubes of the form

$$(r_1 - 1)\, 2^{-k} \leqslant x_1 < r_1 2^{-k}, \quad (r_2 - 1)\, 2^{-k} \leqslant x_2 < r_2 2^{-k}, \ldots,$$
$$(r_n - 1)\, 2^{-k} \leqslant x_n < r_n 2^{-k},$$

and where $k \geqslant 0, r_1, r_2, \ldots, r_n$ are integers. These cubes have the special property that, when two have any point in common then one cube is a subset of the other. Now, if G is any open set, each point g of G lies in a cube C of $\mathscr{C}$ that is contained in G, and so in a cube C of $\mathscr{C}$ which is maximal in the sense that it is a cube C of $\mathscr{C}$ contained in G but not contained in any larger cube C' of $\mathscr{C}$ contained in G. Hence G is the union of the maximal cubes of $\mathscr{C}$ contained in G. Clearly, these maximal cubes C in G are disjoint. So

$$G = \bigcup_{i=1}^{\infty} C_i,$$

where we have a disjoint union of cubes C_i of $\mathscr{C}$. As the Borel sets are measurable for both $\mu^{(h)}$ and λ we conclude that

$$\mu^{(n)}(G) = \sum_{i=1}^{\infty} \mu^{(n)}(C_i) = \kappa_n \sum_{i=1}^{\infty} \lambda(C_i) = \kappa_n \lambda(G), \tag{1}$$

for each open set G.

Now consider any $\mathscr{G}_\delta$-set H. If $\mu^{(n)}(H) = \lambda(H) = +\infty$ we have

$$\mu^{(n)}(H) = \kappa_n \lambda(H).$$

If $\mu^{(n)}(H) < +\infty$, we can cover H by a sequence $\{G_i\}$ of open sets with

$$\sum_{i=1}^{\infty} \{d(G_i)\}^n$$

finite. Then, as in an earlier paragraph, we can replace the open sets G_i by open cubes C_i of edge twice the diameter of G_i, and obtain an open cover $\bigcup_{i=1}^{i=\infty} C_i$ of H with

$$\lambda\left(\bigcup_{i=1}^{\infty} C_i\right) < +\infty.$$

On the other hand, if $\lambda(H) < +\infty$, it is immediate that H can be covered by an open set of finite λ-measure. It follows from (1) that such open sets of finite λ-measure have finite $\mu^{(n)}$-measure. Thus, when H is a $\mathscr{G}_\delta$-set with either $\mu^{(n)}(H) < +\infty$, or with $\lambda(H) < +\infty$, we can express H in the form

$$H = \bigcap_{i=1}^{\infty} G_i,$$

where $G_1, G_2, \ldots,$ is a non-increasing sequence of measurable open sets of finite measure, both for $\mu^{(n)}$ and for λ. Hence using (1) and theorem 8 we have

$$\mu^{(n)}(H) = \inf_i \mu^{(n)}(G_i) = \kappa_n \inf_i \lambda(G_i) = \kappa_n \lambda(H). \tag{2}$$

So this result (2) holds for all $\mathscr{G}_\delta$-sets H.

As $\mu^{(n)}$ and λ are $\mathscr{G}_\delta$-regular measures, if E is any set, we can choose $\mathscr{G}_\delta$-sets H_1, H_2 with

$$E \subset H_1, \qquad \mu^{(n)}(E) = \mu^{(n)}(H_1),$$
$$E \subset H_2, \qquad \lambda(E) = \lambda(H_2).$$

This ensures that

$$\mu^{(n)}(E) = \mu^{(n)}(H_1 \cap H_2) = \kappa_n \lambda(H_1 \cap H_2) = \kappa_n \lambda(E),$$

as required.

Remarks. When $n = 1$ it is easy to see that $\kappa_1 = 1$. The determination of κ_n for $n \geqslant 2$ is more difficult. Using the theory of symmetrization, any convex set of diameter d in E_n has Lebesgue measure not exceeding the Lebesgue measure

$$\frac{\pi^{\frac{1}{2}n}}{\Gamma(1+\frac{1}{2}n)}(\tfrac{1}{2}d)^n$$

of a sphere of diameter d (see, for example, H. G. Eggleston, 1958a, pp. 106–7). So any set of diameter d in E_n is contained in a convex set of Lebesgue measure not exceeding

$$\frac{(\frac{1}{4}\pi)^{\frac{1}{2}n}}{\Gamma(1+\frac{1}{2}n)}d^n.$$

It follows by the argument used in the above proof that, if $\{S_i\}$ is any sequence of sets covering the unit cube C_0, we have

$$\sum_{i=1}^{\infty}\{d(S_i)\}^n\frac{(\frac{1}{4}\pi)^{\frac{1}{2}n}}{\Gamma(1+\frac{1}{2}n)} \geqslant \lambda(C_0),$$

so that
$$\mu^{(n)}(C_0) \geqslant \left(\frac{4}{\pi}\right)^{\frac{1}{2}n}\Gamma(1+\tfrac{1}{2}n)\,\lambda(C_0).$$

and
$$\kappa_n \geqslant \left(\frac{4}{\pi}\right)^{\frac{1}{2}n}\Gamma(1+\tfrac{1}{2}n).$$

On the other hand, if $\delta > 0$ is given, we can by the standard Vitali theorem argument find an infinite sequence $\{S_i\}$ of disjoint spheres of diameter less than δ contained in C_0 with

$$\sum_{i=1}^{\infty}\lambda(S_i) = \lambda(C_0), \qquad \lambda\left(C_0\setminus\bigcup_{i=1}^{\infty}S_i\right) = 0.$$

Then
$$\mu^{(n)}\left(C_0\setminus\bigcup_{i=1}^{\infty}S_i\right) = 0.$$

Hence
$$\begin{aligned}\mu_\delta^{(n)}(C_0) &\leqslant \mu_\delta^{(n)}\left(\bigcup_{i=1}^{\infty}S_i\right)+\mu_\delta^{(n)}\left(C_0\setminus\bigcup_{i=1}^{\infty}S_i\right)\\ &= \mu_\delta^{(n)}\left(\bigcup_{i=1}^{\infty}S_i\right)\\ &\leqslant \sum_{i=1}^{\infty}\{d(S_i)\}^n\\ &= \sum_{i=1}^{\infty}\left(\frac{4}{\pi}\right)^{\frac{1}{2}n}\Gamma(1+\tfrac{1}{2}n)\,\lambda(S_i)\\ &= \left(\frac{4}{\pi}\right)^{\frac{1}{2}n}\Gamma(1+\tfrac{1}{2}n)\,\lambda(C_0).\end{aligned}$$

Consequently
$$\mu^{(n)}(C_0) \leqslant \left(\frac{4}{\pi}\right)^{\frac{1}{2}n}\Gamma(1+\tfrac{1}{2}n),$$

so that $$\kappa_n \leqslant (4/n)^{\frac{1}{2}n}\Gamma(1+\tfrac{1}{2}n).$$
Thus $$\kappa_n = (4/n)^{\frac{1}{2}n}\Gamma(1+\tfrac{1}{2}n),$$
as required.

Perhaps the most appropriate way of describing the relationship between λ and $\mu^{(n)}$ is to say that, while λ assigns unit measure to the cube of unit side, $\mu^{(n)}$ assigns unit measure to the sphere of unit diameter.

2.3. We complete this section with the following mapping theorem.

Theorem 31. *Let Ω be a metric space with metric ρ. Let f be a one–one map of a bounded open set G in E_r into Ω. Suppose that for some positive constants c, C we have*
$$c|\mathbf{x}_2-\mathbf{x}_1| \leqslant \rho(f(\mathbf{x}_1),f(\mathbf{x}_2)) \leqslant C|\mathbf{x}_2-\mathbf{x}_1|,$$
for all points x_1, x_2 of G. Then $f(G)$ has finite positive (r)-measure in Ω.

Proof. As G has a finite positive Lebesgue measure in E_r it has a finite positive (r)-measure in E, by theorem 30.

Now we take $g(t) = Ct$ for $t \geqslant 0$, in theorem 29. As G has finite $\mu^{(r)}$-measure, it follows from theorem 29 that $f(G)$ has finite μ^{ϕ}-measure with
$$\phi(t) = (t/C)^r.$$
Hence, $f(G)$ has finite $\mu^{(r)}$-measure.

If $f(G)$ had zero $\mu^{(r)}$-measure in Ω, we could apply theorem 29 in the reverse direction and deduce that G had zero μ^{ψ}-measure with
$$\psi(t) = (tc)^r,$$
and so zero (r)-measure. Consequently $f(G)$ must have positive (r)-measure.

We remark that, if we use (r)-measure in E_n, then it has many of the desiderata of a concept of an r-dimensional surface area. It is clearly invariant under translations and rotations. It has appropriate additivity properties. It assigns a finite positive value to every sufficiently smooth image of any bounded open set in E_r. See also § 1 of Chapter 3.

§ 3 Existence theorems

We establish a variety of existence theorems in this section. We first show that a σ-compact set E in a metric space has $\mu^h(E) = 0$ for some h in $\mathscr{H}$. A natural comparison to this result would be to show that an uncountable complete separable metric space Ω has $\mu^h(\Omega) > 0$ for some h in $\mathscr{H}_0$; in fact we prove that any such Ω contains a perfect subset C with $0 < \mu^h(C) < +\infty$ for some h in $\mathscr{H}_0$. As a third existence theorem we show that if h is a given function of $\mathscr{H}_0$ there is a compact metric Ω with $0 < \mu^h(\Omega) < +\infty$. We leave for consideration in §7 the more difficult existence problem when h and Ω are given and a subset C is sought with $0 < \mu^h(C) < +\infty$. We conclude the section with a result, based on the continuum hypothesis, showing that there can be uncountable sets which are of h-measure zero for all h in $\mathscr{H}_0$.

3.1. Before we formulate any of the existence theorems it will be convenient to have a criterion for a set E to have h-measure zero.

Theorem 32. *A set E has zero h-measure, if, and only if, it is possible to choose a sequence $\{E_i\}$ of sets, with $\sum_{i=1}^{i=\infty} h(E_i)$ finite, so that each point E belongs to infinitely many of the sets E_i.*

Proof. Suppose $\mu^h(E) = 0$. Then, by the definition of μ^h, we can, for each $n \geqslant 1$, choose a sequence of sets $\{E_i^{(n)}\}$ with

$$E \subset \bigcup_{i=1}^{\infty} E_i^{(n)}, \qquad \sum_{i=1}^{\infty} h(E_i^{(n)}) < 2^{-n}.$$

Now, it is clear that rearranging the sets $E_i^{(n)}$ $(i, n = 1, 2, \ldots)$, as a single sequence we satisfy the requirements of the theorem, as

$$\sum_{i=1}^{\infty} \sum_{n=1}^{\infty} h(E_i^{(n)}) < 1,$$

and each point of E belongs to infinitely many of the sets $E_i^{(n)}$.

Now suppose we have sets $\{E_i\}$ with $\sum_{i=1}^{i=\infty} h(E_i)$ finite and that each point of E belongs to infinitely many of the sets. Let $\epsilon > 0$ be given. Let $\delta > 0$ be given. Then $h(\delta) > 0$. Hence $d(E_i) \geqslant \delta$, for at most a finite number of values of i. So we can choose an integer N so that $d(E_i) < \delta$ for all $i \geqslant N$. We may suppose, in addition, that N is so large that

$$\sum_{i=N}^{\infty} h(E_i) < \epsilon.$$

Now, as each point of E belongs to infinitely many of the sets E_i, we have

$$E \subset \bigcup_{i=N}^{\infty} E_i, \qquad d(E_i) < \delta \quad \text{for} \quad i \geqslant N, \qquad \sum_{i=N}^{\infty} h(E_i) < \epsilon.$$

Hence $$\mu_\delta^h(E) < \epsilon.$$

Thus $\mu^h(E) = 0$ as required.

3.2. We shall be concerned to use h-measures to estimate the 'sizes' of various sets. A set that cannot be covered by any countable sequence of sets of diameter δ, for some $\delta > 0$, is so large that it automatically has infinite μ^h-measure, no matter what h in $\mathscr{H}$ is used. We first show that σ-compact sets are not so large as this.

Theorem 33. *Let E be a σ-compact set in a metric space Ω. Then there is an h in $\mathscr{H}$ for which $\mu^h(E) = 0$.*

Proof. As E is σ-compact we can write

$$E = \bigcup_{i=1}^{\infty} K_i,$$

where each set K_i is compact. Write

$$F_i = \bigcup_{j=1}^{i} K_j,$$

for $i = 1, 2, \ldots$. Then each set F_i is compact,

$$F_1 \subset F_2 \subset F_3 \subset \ldots,$$

and

$$E = \bigcup_{i=1}^{\infty} F_i.$$

For each i, the system of open spheres

$$S(x, 1/i) \quad (x \in F_i),$$

where we define $S(x, r)$ to be the sphere of points y with

$$\rho(x, y) < r,$$

forms an open cover of the compact set F_i. Let

$$S(x_{ij}, 1/i) \quad (j = 1, 2, \ldots, J_i),$$

be a finite subcover of F_i by such spheres. Let $S_1, S_2, \ldots$ be an enumeration of these spheres in their natural order, taking first those associated with F_1, then those associated with F_2, and so on. Then

$$d(S_k) \leqslant 2/i,$$

whenever

$$k > J_1 + J_2 + \ldots + J_{i-1}.$$

Now we can choose h to be a continuous and monotonic increasing function satisfying the conditions

$$h(0) = 0,$$

$$h(2/i) = \{J_1 + J_2 + \ldots + J_i\}^{-2} \quad (i = 1, 2, 3, \ldots).$$

Now given any $k > J_1$, there will be an i with

$$J_1+J_2+\ldots+J_{i-1} < k \leqslant J_1+J_2+\ldots+J_i,$$

and with this value of i,

$$h(S_k) = h(d(S_k)) \leqslant h(2/i) = \{J_1+J_2+\ldots+J_i\}^{-2} \leqslant k^{-2}.$$

Hence the series $\sum_{k=1}^{k=\infty} h(S_k)$ converges. But each point of E belongs to all sets F_i from some point onwards and so to infinitely many sets S_k. Thus the criterion of theorem 32 is satisfied and $\mu^h(E) = 0$.

3.3. Our next objective will be to prove a result in the other direction showing that provided a space is 'not too small' it will contain subsets having positive h-measure for some h in $\mathscr{H}_0$. For this purpose we need the Cantor–Bendixson theorem; we prove it for the convenience of the reader. Its statement and proof requires some definitions.

***Definition* 17.** *A class $\mathscr{G}^*$ of open sets is said to form a base for the open sets of a topological space Ω, if each open set of Ω is a union of sets from $\mathscr{G}^*$.*

***Definition* 18.** *A set is dense in itself if each of the open sets that meets it at all, meets it in an infinite set.*

***Definition* 19.** *A set is perfect if it is closed, non-empty and dense in itself.*

***Definition* 20.** *A point c is said to be a point of condensation of a set S if each open set containing c meets S in an uncountable set.*

***Theorem* 34.** (Cantor–Bendixson). *Let Ω be a topological space with a countable base for its open sets. Then any uncountable closed set in Ω contains a perfect subset.*

Proof. Let F be an uncountable closed subset of Ω. Choose a countable base for the open sets of Ω, name the sets of the base

$$U_1, U_2, \ldots.$$

Let $V_1, V_2, \ldots$ be those of the sets $U_1, U_2, \ldots$ that meet F in a countable set. So the sequence $V_1, V_2, \ldots$ may be finite or infinite. Write

$$P = F\backslash\{\bigcup_i V_i\},$$

Then P is closed and $\quad F = P \cup \{\bigcup_i V_i \cap F\}.$

As F is uncountable, while $\bigcup_i V_i \cap F$

is countable, it follows that P is uncountable and so certainly non-empty.

It remains to prove that P is dense in itself; it is convenient to prove rather more, that each point of P is a point of condensation of P. Let p be any point of P. Let G be any open set containing p. Then G is the union of those elements of the base that are contained within G. So for some j we have

$$p \in U_j \subset G.$$

If we had $U_j = V_i$ for some i, we would have

$$p \notin F \backslash \{\bigcup_i V_i\} = P.$$

By the choice of the sets V_i it follows that $F \cap U_j$ is uncountable. As P differs from F by merely a countable set, it follows that $P \cap U_j$ is uncountable. As $P \cap U_j \subset P \cap G$, it follows that $P \cap G$ is uncountable. Thus each point p of P is a point of condensation of P. So P is certainly dense in itself.

***Corollary* 1.** *The points of condensation of F form a perfect set P and are points of condensation of P.*

Proof. In the construction used in the proof of the theorem, a point f of condensation of F belongs to no set V_i and so remains in P and becomes a point of condensation of P. Each point of P is a point of condensation of P and *a fortiori* of F, and so arises in this way from a point of condensation of F.

***Corollary* 2.** *Any uncountable set in Ω contains an uncountable subset that is dense in itself.*

Proof. Let F be an uncountable subset of Ω. Proceed as in the proof of the theorem omitting the remark that P is closed. The result follows.

3.4. We can now prove a result due essentially to A. S. Besicovitch (1934); the proof is due to C. A. Rogers and M. Sion (1963) but is based on familiar ideas.

***Definition* 21.** *A metric space Ω is said to be complete, if each sequence $\{x_n\}$ of points, that satisfies the Cauchy condition for convergence—'for each $\epsilon > 0$, there is an integer M_ϵ such that for all m, n with $m > n \geqslant M_\epsilon$ we have $\rho(x_m, x_n) < \epsilon$'—converges to some point of the space.*

***Definition* 22.** *A metric space is called separable if its open sets have a countable base.*

***Theorem* 35.** *Let Ω be an uncountable complete separable metric space. Then there is a compact perfect subset C of Ω and a function h of $\mathscr{H}_0$ with*

$$0 < \mu^h(C) < +\infty. \tag{1}$$

Proof. As a metric space is always a topological space we can use the Cantor–Bendixson theorem to provide a perfect subset P of Ω.

Our aim is to choose in the perfect set P a perfect subset C in a way that facilitates the construction of a function h of $\mathscr{H}_0$ satisfying (1). The construction has many points in common with the construction of the Cantor ternary set.

As P is perfect it is non-empty and we can choose point $c(0)$ in P. Choose any positive number r_0. As P is dense in itself we can choose a second point $c(1)$ in P with

$$0 < \rho(c(0), c(1)) < \tfrac{1}{4}r_0.$$

Suppose, for some integer $k \geqslant 1$, we have chosen families of distinct points

$$c(\mathbf{i}|r) = c(i_1, i_2, \ldots, i_r) \quad (i_1, i_2, \ldots, i_r = 0 \quad \text{or} \quad 1),$$

for $r = 1, 2, \ldots, k$, satisfying

$$0 < \rho(c(i_1, i_2, \ldots, i_r, 1),\ c(i_1, i_2, \ldots, i_r, 0)) < \tfrac{1}{4}r_r$$

for $0 \leqslant r < k$, with
$$r_r = \min_{\mathbf{i}|r \neq \mathbf{j}|r} \rho(c(\mathbf{i}|r),\ c(\mathbf{j}|r)), \tag{2}$$
for $1 \leqslant r \leqslant k$.

Then we take $\quad c(i_1, i_2, \ldots, i_k, 0) = c(i_1, i_2, \ldots, i_k),$

for $i_1, i_2, \ldots, i_k = 0$ or 1, and using the denseness of P in itself, we choose distinct points

$$c(i_1, i_2, \ldots, i_k, 1)$$

in P with
$$0 < \rho(c(i_1, i_2, \ldots, i_k, 1),\ c(i_1, i_2, \ldots, i_k, 0)) < \tfrac{1}{4}r_k. \tag{3}$$

Thus
$$r_{k+1} = \min_{\mathbf{i}|k+1 \neq \mathbf{j}|k+1} \rho(c(\mathbf{i}|k+1),\ c(\mathbf{j}|k+1)) > 0.$$

By (3) we also have
$$r_{k+1} < \tfrac{1}{4}r_k. \tag{4}$$
It is now clear that points

$$c(\mathbf{i}|k) \quad (i_1, i_2, \ldots, i_k = 0 \quad \text{or} \quad 1),$$

can be chosen inductively in this way for $k = 1, 2, 3, \ldots$.

For each infinite sequence $\mathbf{i} = i_1, i_2, \ldots$ of 0's and 1's we study the sequence

$$c(\mathbf{i}|1),\ c(\mathbf{i}|2),\ c(\mathbf{i}|3), \ldots.$$

If $t > k$, we have

$$\begin{aligned}\rho(c(\mathbf{i}|k), c(\mathbf{i}|t)) &\leqslant \sum_{r=0}^{t-k-1} \rho(c(\mathbf{i}|k+r), c(\mathbf{i}|k+r+1)) \\ &= \sum_{r=0}^{t-k-1} \rho(c(\mathbf{i}|k+r, 0), c(\mathbf{i}|k+r+1)) \\ &\leqslant \sum_{r=0}^{t-k-1} \tfrac{1}{4} r_{k+r} \\ &\leqslant \sum_{r=0}^{t-k-1} \tfrac{1}{4}(\tfrac{1}{4})^r r_k \\ &< \tfrac{1}{3} r_k \leqslant \tfrac{1}{3}(\tfrac{1}{4})^{k-1} r_1, \end{aligned} \tag{5}$$

on using (4). So the sequence is a Cauchy sequence and so converges to a point $c(\mathbf{i})$ say of Ω. Further, on letting $t \to \infty$ in (5) we obtain

$$\rho(c(\mathbf{i}|k), c(\mathbf{i})) \leqslant \tfrac{1}{3} r_k. \tag{6}$$

We take C to be the set of all these points $c(\mathbf{i})$.

Now consider two different sequences $\mathbf{i}, \mathbf{j}$ of 0's and 1's. There will be a unique integer k with

$$\mathbf{i}|k-1 = \mathbf{j}|k-1, \qquad i_k \neq j_k. \tag{7}$$

Then, using (2) and (6),

$$\begin{aligned}\rho(c(\mathbf{i}), c(\mathbf{j})) &\geqslant \rho(c(\mathbf{i}|k), c(\mathbf{j}|k)) - \rho(c(\mathbf{i}), (\mathbf{i}|k)) - \rho(c(\mathbf{j}), c(\mathbf{j}|k)) \\ &\geqslant r_k - \tfrac{1}{3} r_k - \tfrac{1}{3} r_k = \tfrac{1}{3} r_k. \end{aligned} \tag{8}$$

But, also using (6) and (7)

$$\begin{aligned}\rho(c(\mathbf{i}), c(\mathbf{j})) &\leqslant \rho(c(\mathbf{i}), c(\mathbf{i}|k-1)) + \rho(c(\mathbf{j}), c(\mathbf{j}|k-1)) \\ &\leqslant \tfrac{1}{3} r_{k-1} + \tfrac{1}{3} r_{k-1} = \tfrac{2}{3} r_{k-1}. \end{aligned} \tag{9}$$

These inequalities (8) and (9) show that we have some control over the distances between such points of C.

It is easy to show that C is perfect. If $c \in C$, we have $c = c(\mathbf{i})$ for some sequence $\mathbf{i}$. By taking

$$i_t^{(k)} = i_t \quad (t \neq k),$$

$$i_k^{(k)} = 1 - i_k,$$

we obtain a sequence of distinct vectors

$$\mathbf{i}^{(k)} = i_1^{(k)}, i_2^{(k)}, \ldots \quad (k = 1, 2, \ldots).$$

The corresponding points

$$c(\mathbf{i}^{(k)}) \quad (k = 1, 2, \ldots),$$

form a sequence of distinct points converging to $c = c(\mathbf{i})$. Thus C is dense in itself.

To prove that C is compact we consider any sequence $c_1, c_2, \ldots$, of points of C. Suppose that

$$c_k = c(\mathbf{i}^{(k)})$$

for a sequence $\mathbf{i}^{(k)}$ of 0's and 1's for $k = 1, 2, \ldots$. Now we can choose a sequence $\mathbf{j} = j_1, j_2, \ldots$ of 0's and 1's and a sequence of infinite subsequences $R_1, R_2, \ldots$ of the natural numbers, inductively, so that

$$R_1 \supset R_2 \supset R_3 \supset \ldots,$$

$$i_s^{(k)} = j_s \quad (\text{for} \quad k \in R_s, s = 1, 2, \ldots).$$

Here j_1 is chosen so that

$$i_1^{(k)} = j_1$$

for infinitely many values of k and R_1 is taken to be the set of integers k satisfying this condition. Then, when R_s has been chosen, j_{s+1} is chosen so that

$$i_{s+1}^{(k)} = j_{s+1},$$

for infinitely many integers k in R_s and R_{s+1} is chosen to be the set of these integers. Now take R to be the diagonal sequence whose first element is first element n_1 of R_1 and whose $(s+1)$-st element n_{s+1} is the first element in R_{s+1} exceeding n_s for $s = 1, 2, \ldots$. Then for k in R exceeding n_s we have

$$k \in R_s \subset R_{s-1} \subset \ldots \subset R_2 \subset R_1,$$

and

$$\mathbf{i}^{(k)}|s = \mathbf{j}|s,$$

so that, using (9),

$$\rho(c_k, c(\mathbf{j})) = \rho(c(\mathbf{i}^{(k)}), c(\mathbf{j})) \leqslant \tfrac{2}{3}r_s.$$

Hence, c_k converges to the point $c(\mathbf{j})$ of C as k tends to infinity through the subsequence R. Thus C is sequentially compact and *a fortiori* closed. As the space is separable, it follows that C is compact. This completes the proof that C is a compact perfect set.

We now choose a function h of $\mathscr{H}_0$ and show that $\mu^h(C)$ is finite and positive. We recall that the sequence $r_1, r_2, \ldots$ of positive numbers satisfies the condition

$$r_{k+1} < \tfrac{1}{4}r_k \quad (k = 1, 2, \ldots).$$

So there are certainly functions h in $\mathscr{H}_0$ satisfying the conditions

$$h(0) = 0,$$

$$h(t) = 2^{-k} \quad (\text{for } \tfrac{1}{3}r_k \leqslant t \leqslant \tfrac{2}{3}r_k, k = 1, 2, \ldots).$$

We choose any such function.

Let $\Sigma(\mathbf{i}|k)$ denote the closed sphere of all points x with

$$\rho(x, c(\mathbf{i}|k)) \leqslant \tfrac{1}{3}r_k.$$

Then
$$d(\Sigma(\mathbf{i}|k)) \leqslant \tfrac{2}{3}r_k,$$
and
$$h(\Sigma(\mathbf{i}|k)) \leqslant h(\tfrac{2}{3}r_k) = 2^{-k}.$$

But, by (6), each point $c(\mathbf{i})$ of C lies in the corresponding sphere

$$\Sigma(\mathbf{i}|k).$$

As there are just 2^k of these spheres, we have

$$\mu_\delta^h(C) \leqslant \sum_{\mathbf{i}|k} h(\Sigma(\mathbf{i}|k)) \leqslant 1,$$

for $\delta > \frac{2}{3}r_k$. Consequently $\mu^h(C) \leqslant 1$.

Now suppose that $\{G_i\}$ is any sequence of open sets covering C. As C is compact we can choose an integer N so large that

$$C \subset \bigcup_{i=1}^{N} G_i.$$

After renaming the sets and changing N, we may suppose that

$$G_i \cap C \neq \varnothing \quad (i = 1, 2, \ldots, N).$$

As each point of C is a limit point of points of the form $c(\mathbf{i}|k)$, we can choose an integer k^* so large that each set G_i, $i = 1, 2, \ldots, N$ contains at least *two* of the points

$$c(\mathbf{i}|k^*).$$

Let g_i be the number of these points in G_i and choose h_i to be the integer with

$$2^{h_i} < g_i \leqslant 2^{h_i+1} \quad (i = 1, 2, \ldots, N).$$

Then there are more than 2^{h_i} of the points $c(\mathbf{i}|k^*)$ in G_i, but at most 2^{h_i} of these k^*-vectors $\mathbf{i}|k^*$ can agree in their first $k^* - h_i$ components. Hence we can choose $\mathbf{i}$ and $\mathbf{j}$ with

$$c(\mathbf{i}|k^*) \in G_i, \qquad c(\mathbf{j}|k^*) \in G_i,$$
$$\mathbf{i}|(k^* - h_i) \neq \mathbf{j}|(k^* - h_i).$$

By (8) we deduce that
$$d(G_i) \geqslant \tfrac{1}{3} r_{k^*-h_i},$$
so that
$$h(G_i) \geqslant h(\tfrac{1}{3}r_{k^*-h_i}) = 2^{-k^*+h_i} \geqslant \tfrac{1}{2}g_i 2^{-k^*}.$$

As the total number of points $c(\mathbf{i}|k^*)$ covered by the sets G_i is 2^{k^*} we conclude that

$$\sum_{i=1}^{\infty} h(G_i) \geqslant \sum_{i=1}^{N} h(G_i) \geqslant \tfrac{1}{2}\sum_{i=1}^{N} g_i 2^{-k^*} \geqslant \tfrac{1}{2}.$$

Consequently
$$\mu^h(C) \geqslant \tfrac{1}{2},$$
and
$$\tfrac{1}{2} \leqslant \mu^h(C) \leqslant 1,$$
as required.

Remark. The way we have organized this proof has to some extent obscured the close connection between our construction and the standard method of constructing the Cantor ternary set. But, if we write
$$C_k = \bigcup_{\mathbf{i}|k} \Sigma(\mathbf{i}|k) \quad (k = 1, 2, \ldots).$$
where $\Sigma(\mathbf{i}|k)$ is the closed sphere introduced above, it is easy to verify that
$$C = \bigcap_{k=1}^{\infty} C_k.$$
Now the closed spheres $\Sigma(\mathbf{i}|k)$ can be regarded as playing the same role in the construction of C as is played by the closed intervals involved in the usual construction of the Cantor ternary set.

Corollary 1. *Let K be an uncountable compact set in a metric space Ω. Then there is a perfect subset C of K and a function h in $\mathscr{H}_0$ with*
$$0 < \mu^h(C) < +\infty.$$

Proof. As K is a complete separable metric space under the metric of Ω, the result follows.

Corollary 2. *If Ω is a complete separable metric space and $\mu^h(K) = 0$ for each h in $\mathscr{H}_0$ and each compact K in Ω, then Ω is countable.*

Corollary 3. *If K is a compact set in a metric space and $\mu^h(K) = 0$ for all h in $\mathscr{H}_0$, then K is countable.*

Corollary 4. *If A is an analytic set (see the proof below) in a metric space and $\mu^h(A) = 0$ for all h in $\mathscr{H}_0$, then A is countable.*

Proof in outline. Let A be an uncountable analytic set in Ω, that is, an uncountable set that is the continuous image of the irrationals between 0 and 1. Then A contains an uncountable compact set K (see W. Sierpiński (1956) or Kuratowski (1966)). By the theorem, there is an h in $\mathscr{H}_0$ with $0 < \mu^h(K) \leqslant \mu^h(A)$, contrary to our assumptions. Note that, in a complete separable metric space, each Borel set is an analytic set.

Corollary 5. *Let Ω be a complete separable metric space. Then there is a Borel regular metric measure μ on Ω, that assigns: zero measure to each point of Ω; positive measure to each uncountable open set of Ω; and finite measure to Ω.*

Proof in outline. Choose a closed basis $\{V_i\}_{i=1}^{i=\infty}$ for the open sets of Ω. For each uncountable V_i, choose an h_i in $\mathscr{H}_0$ and a perfect set C_i in V_i with $\mu^{h_i}(C_i) = 2^{-i}$. Define μ by

$$\mu(E) = \Sigma' \mu^{h_i}(C_i \cap E),$$

the sum being taken over the integers i for which V_i is uncountable.

3.5. This last theorem naturally raises three questions. If we are given the space Ω and a compact subset C can we hope to choose a function h of $\mathscr{H}_0$ with

$$0 < \mu^h(C) < +\infty?$$

The answer here turns out to be 'No! It is unusual for a set to have such a function h associated with it.' The second question is: if we are given h in $\mathscr{H}_0$ with $\mu^h(\Omega) = +\infty$, will Ω contain a subset of finite positive μ^h-measure? A. S. Besicovitch has answered this question in the affirmative in Euclidean space; we discuss his work and later work of R. O. Davies and of D. G. Larman on this problem in §§ 6 and 7 below.

The third question is: if we are given h in $\mathscr{H}_0$ can we find a compact metric space Ω having finite positive μ^h-measure? This we can do without much difficulty.

Theorem 36. *Let h be a function of $\mathscr{H}_0$. Then there is a compact metric space Ω with*

$$0 < \mu^h(\Omega) < +\infty.$$

Proof. It will be convenient to take the points of Ω to be the points of a certain compact subset of the closed interval $I_0 = [0, 1]$ of the real line; it will be necessary to define the metric in Ω so that it is quite different from the usual metric on the real line.

For convenience, we write

$$\nu(t) = \left[\frac{1}{h(t)}\right] \quad (t > 0),$$

the square bracket being the 'integral-part' function used in number theory. Since

$$h(t) \to 0 \quad \text{as} \quad t \to 0,$$

we may choose $t_0 > 0$ so that

$$\nu(t_0) \geqslant 2,$$

and we can then choose $t_1, t_2, \ldots$ inductively so that

$$h(t_r) \leqslant r^{-2} \qquad (r = 1, 2, \ldots), \tag{10}$$

$$\nu(t_r) \geqslant (r^2+1)\,\nu(t_{r-1}) \qquad (r = 1, 2, \ldots). \tag{11}$$

Write $n(r) = \nu(t_r)$ and

$$\lambda_r = \left[\frac{\nu(t_r)}{\nu(t_{r-1})}\right] = \left[\frac{n(r)}{n(r-1)}\right] \quad (r = 1, 2, \ldots),$$

then

$$\lambda_r \geqslant r^2+1, \tag{12}$$

and

$$n(r) = \lambda_r n(r-1) + \mu_r \quad (r = 1, 2, \ldots), \tag{13}$$

where $0 \leqslant \mu_r < n(r-1)$.

We make an inductive choice of systems of closed sub-intervals of I_0. For $r = 1, 2, \ldots$, there will be $n(r)$ disjoint closed sub-intervals

$$J_i^{(r)} \quad (i = 1, 2, \ldots, n(1)). \tag{14}$$

The system $J_i^{(1)}$ $(i = 1, 2, \ldots, n(1))$ of disjoint closed sub-intervals of equal length is chosen arbitrarily. When the system (14) has been chosen for a particular positive integer r, the disjoint system of closed sub-intervals of equal length

$$J_i^{(r+1)} \quad (i = 1, 2, \ldots, n(r+1)),$$

is chosen arbitrarily subject to the conditions that μ_{r+1} of the intervals $J_i^{(r)}$ contain precisely $\lambda_{r+1}+1$ of the intervals $J_i^{(r+1)}$ and that the remaining $n(r)-\mu_{r+1}$ of the intervals $J_i^{(r)}$ contain precisely λ_{r+1} of the intervals $J_i^{(r+1)}$. By (13), this accounts for all the intervals $J_i^{(r+1)}$, so that each of these intervals lies in one of the intervals $J_i^{(r)}$.

Having made such a choice of systems of intervals, we take Ω to be the set

$$\bigcap_{r=1}^{\infty} \bigcup_{i=1}^{n(r)} J_i^{(r)}$$

of real numbers. As a set of real numbers, Ω is compact and perfect.

We now define a metric ρ in Ω. We take $\rho(x, x) = 0$, as we must, for all x in Ω. If x, y are distinct points of Ω, they are real numbers at a positive distance apart. For each r, both x and y belong to

$$\bigcup_{i=1}^{n(r)} J_i^{(r)}.$$

These sub-intervals are all of equal length, tending to zero as r tends to infinity. So when r is sufficiently large x and y belong to different intervals $J_i^{(r)}$. Let $r(x, y)$ be the largest integer r for which x and y lie in the same sub-interval $J_i^{(r)}$, with the convention that there is just the one sub-interval $J_1^{(0)} = I_0$ of zero order, and define

$$\rho(x, y) = t_{r(x, y)}.$$

It is easy to verify that the function $\rho(x, y)$ defined on Ω in this way satisfies the conditions for a metric.

Now Ω has two topologies, the relative topology as a subset of the real line, and the metric topology defined by the metric ρ. We show that these two topologies coincide. First suppose that G is a subset of Ω that is open under its relative topology. Then $G = \Omega \cap H$ where H is some open subset of $[0, 1]$. Let x be any point of G. Then $x \in G \subset H$. So for some $\delta > 0$, all points y of $[0, 1]$ with $|x-y| < \delta$ belong to H and so all points y of Ω with $|x-y| < \delta$ belong to G. Now we can choose r so large that all the intervals $J_i^{(r)}$ have length less than δ. Then all points y of Ω with

$$\rho(x, y) \leqslant t_r$$

lie in the same interval $J_i^{(r)}$ as x, and so satisfy $|x-y| < \delta$ and belong to G. Thus each point of G is an inner point of G under the metric ρ and so G is open under ρ. Now suppose that G is a set of Ω that is open under the metric topology. Write

$$H = G \cup \{[0, 1] \backslash \Omega\}.$$

Then $G = \Omega \cap H$ and G will be open in the relative topology provided H is open as a subset of $[0, 1]$. Consider any point x of H. If

$$x \in \{[0, 1] \backslash \Omega\},$$

then x is an inner point of H, as Ω is a closed subset of $[0, 1]$. If $x \in G$, we can choose r so large that all points y of Ω with

$$\rho(x, y) \leqslant t_r$$

lie in G. Let x lie in a particular interval $J_{i*}^{(r)}$ of order r. Then all points y of $[0, 1]$ sufficiently close to x lie either in $[0, 1] \backslash \Omega$ or in $J_{i*}^{(r)}$. In the first case

$$y \in [0, 1] \backslash \Omega \subset H,$$

and in the second case $\rho(x, y) \leqslant t_r$ and

$$y \in G \subset H.$$

Hence H is open. This shows that the two topologies on Ω coincide.

Now Ω being a closed subset of $[0, 1]$ is a compact metric space under its relative topology. Hence Ω is a compact metric space under its metric ρ.

It is easy to check that Ω has finite μ^h-measure. Ω is covered by the finite system of sets

$$\Omega \cap J_i^{(r)} \quad (i = 1, 2, \ldots, n(r)),$$

for each fixed r. But the diameter of $\Omega \cap J_i^{(r)}$ is t_r as there are points of this set lying in different intervals $J_i^{(r+1)}$. So, provided $0 < t_r \leqslant \delta$, we have

$$\mu_\delta^h(\Omega) \leqslant \sum_{i=1}^{n(r)} h(\Omega \cap J_i^{(r)}) = n(r)\, h(t_r) \leqslant \nu(t_r)\, h(t_r) = \left[\frac{1}{h(t_r)}\right] h(t_r) \leqslant 1.$$

For given $\delta > 0$, the condition $0 < t_r \leqslant \delta$ is satisfied as soon as r is sufficiently large. Hence $\mu_\delta^h(\Omega) \leqslant 1$, for all $\delta > 0$, and so $\mu^h(\Omega) \leqslant 1$.

To prove that Ω has positive μ^h-measure we use a standard technique. Suppose a positive integer R is given. Choose $\delta = \delta(R)$ so small that $0 < \delta < t_R$. Consider any cover of Ω by a sequence $\{G_j\}$ of open sets with

$$\Omega \subset \bigcup G_j, \qquad d(G_j) \leqslant \delta.$$

We seek a lower bound for the sum

$$\sum_{j=1}^{\infty} h(G_j).$$

As Ω is compact we may confine our attention to a finite system of sets, say the sets $G_1, G_2, \ldots, G_N$ with

$$\Omega \subset \bigcup_{j=1}^{\infty} G_j, \qquad d(G_j) \leqslant \delta.$$

As the only possible positive distances between points of Ω are the numbers in the strictly decreasing sequence

$$t_1, t_2, \ldots,$$

the diameter of each set in Ω is an attained distance between two points of the set. As Ω is a perfect subset of $[0,1]$ and G_j is open in Ω, G_j has positive diameter. Suppose G_j has diameter $t_{r(j)}$. Then there will be two points of G_j in the same interval $J_i^{(r(j))}$ lying in different intervals $J_i^{(r(j)+1)}$. Further all points of G_j must lie in the one interval $J_{i(j)}^{(r(j))}$. Thus

$$G_j \subset J_{i(j)}^{(r(j))} \quad \text{and} \quad d(G_j) = d(J_{i(j)}^{(r(j))}).$$

So it will suffice for our purpose to obtain a lower bound for

$$\sum_{i=1}^{N} h(J_{i(j)}^{(r(j))}),$$

over all sequences $\quad J_{i(j)}^{(r(j))} \quad (j = 1, 2, \ldots, N),$

subject to the conditions

$$\Omega \subset \bigcup_{i=1}^{N} J_{i(j)}^{(r(j))} \quad (d(J_{i(j)}^{(r(j))}) \leqslant \delta).$$

The condition $d(J_{i(j)}^{(r(j))}) \leqslant \delta$ we replace by the weaker condition

$$d(J_{i(j)}^{(r(j))}) \leqslant t_R,$$

which is equivalent to
$$t_{r(j)} \leqslant t_R,$$
or to
$$r(j) \geqslant R.$$

But, if two of the intervals $J_{i(j)}^{(r(j))}$ meet, then one must be contained in the other and so will be redundant in the covering. So we may assume that all the intervals $J_{i(j)}^{(r(j))}$ are disjoint. Write $r = \max_{1 \leqslant j \leqslant N} r(j)$, and let $I_1, I_2, \ldots, I_s$ be the intervals of order r amongst the intervals $J_{i(j)}^{(r(j))}$, $j = 1, 2, \ldots, N$. Clearly these intervals $I_1, I_2, \ldots, I_s$ must be *the* intervals of order r lying in some set $K_1, K_2, \ldots, K_t$ of intervals of order $r-1$. As each interval of order $r-1$ contains either λ_r or $\lambda_r + 1$ intervals of order r, we conclude that $s \geqslant \lambda_r t$ and

$$\sum_{i=1}^{s} h(I_i) \geqslant sh(t_r) \geqslant \lambda_r h(t_r)\, t = \lambda_r \frac{h(t_r)}{h(t_{r-1})} \sum_{i=1}^{t} h(K_i).$$

But
$$\begin{aligned}
\lambda_r \frac{h(t_r)}{h(t_{r-1})} &= (\lambda_r + 1) \frac{1}{1 + (\lambda_r)^{-1}} \frac{h(t_r)}{h(t_{r-1})} \\
&\geqslant \frac{\nu(t_r)}{\nu(t_{r-1})} \frac{1}{1 + r^{-2}} \frac{h(t_r)}{h(t_{r-1})} \\
&= \frac{h(t_r)\,[1/h(t_r)]}{h(t_{r-1})\,[1/h(t_{r-1})]} \frac{1}{1 + r^{-2}} \\
&\geqslant h(t_r) \left\{ \frac{1}{h(t_r)} - 1 \right\} \frac{1}{1 + r^{-2}} \\
&= (1 - h(t_r))/(1 + r^{-2}) \\
&\geqslant (1 - r^{-2})/(1 + r^{-2}) \\
&> (1 - r^{-2})^2,
\end{aligned}$$

on using (12) and (10). Hence

$$\sum_{i=1}^{s} h(I_i) \geqslant (1 - r^{-2})^2 \sum_{i=1}^{t} h(K_i).$$

By replacing the intervals $I_1, \ldots, I_s$ by the intervals $K_1, \ldots, K_t$, we replace the system $J_{i(j)}^{(r(j))}$, $j = 1, 2, \ldots, N$, of disjoint intervals covering Ω and having orders between R and r inclusive, by a similar set of intervals having orders between R and $r-1$, the sum of the h-values of the original system being at least $(1 - r^{-2})^2$ times the sum of h-values of the new system. After $r-R$ replacements of this type we obtain *the* system of intervals $J_i^{(R)}$ $(i = 1, 2, \ldots, n(R))$. Hence

$$\begin{aligned}\sum_{j=1}^{N} h(J_{i(j)}^{(r(j))}) &\geqslant \left\{\prod_{s=R+1}^{r} (1-s^{-2})^2\right\} \sum_{i=1}^{n(R)} h(J_i^{(R)}) \\ &= \left\{\prod_{s=R+1}^{r} (1-s^{-2})^2\right\} n(R)\, h(t_R) \\ &= \left\{\prod_{s=R+1}^{r} (1-s^{-2})^2\right\} [1/h(t_R)]\, h(t_R) \\ &\geqslant \prod_{s=R}^{\infty} (1-s^{-2})^2.\end{aligned}$$

Consequently $$\mu^h(\Omega) \geqslant \prod_{s=R}^{\infty} (1-s^{-2})^2.$$

As this holds for each $R \geqslant 2$, we deduce that $\mu^h(\Omega) \geqslant 1$. Thus $\mu^h(\Omega) = 1$ as required.

Remark. The above proof is a simplified version of a proof due to A. Dvoretzky. Dvoretzky works in n-dimensional Euclidean space and, on the assumption that

$$\liminf_{t \to 0} h(t)/t^n = +\infty,$$

constructs a Cantor-like set with μ^h-measure unity. (See A. Dvoretzky, 1948, and the references quoted there for earlier work.) If h were continuous, with $h(1) > 1$, the proof could be made very much simpler, it would be possible to choose the sequence $\{t_r\}$ so that

$$h(t_r) = 2^{-r} \quad (r = 0, 1, \ldots),$$

to take Ω to be the usual Cantor ternary set, the metric $\rho(x, y)$ being taken to be t_r if x, y belong to the same rth order intervals but to different $(r+1)$st order intervals. We have given the proof in the above refined form as we need the full result in proving theorem 41 below.

3.6. It might be thought that the restriction in corollary 3 to theorem 35 to compact sets is unnecessary. We show later how it can be relaxed in some cases (see §6); but our next result shows that it is not true generally. Originally proved by A. S. Besicovitch (1934), the following proof is due to Dieudonné.

***Theorem* 37.** *Let Ω be a complete separable metric space with no isolated point. On the continuum hypothesis there is an uncountable set C in Ω with $\mu^h(C) = 0$ for all h in $\mathscr{H}_0$.*

This result follows immediately from theorems 38 and 39 stated below. These theorems depend on some definitions.

***Definition* 23.** *A point a in a space Ω is said to be isolated, if the set $\{a\}$ is open.*

***Definition* 24.** *A set N in a space Ω is said to be nowhere dense, if each non-empty open set G of Ω contains a non-empty open subset that does not meet N.*

***Definition* 25.** *A set N in a space Ω is said to be of the first category, if it is a countable union of nowhere dense sets; otherwise it is said to be of the second category.*

***Definition* 26.** *A set C is said to be concentrated on a set D, if each open set containing D contains all points of C with at most a countable number of exceptions.*

***Theorem* 38.** *Let Ω be a complete separable metric space with no isolated point. On the continuum hypothesis there is an uncountable set C in Ω that is concentrated on a countable set D in Ω.*

Proof. As Ω is separable, it follows directly from definition 22 that we can choose a countable base $V_1, V_2, \ldots$ for the open sets. We may suppose that, for each i, $V_i \neq \varnothing$, and we can choose a point d_i in V_i for each i. Take D to be the countable set of all the points

$$d_i \quad (i = 1, 2, \ldots).$$

We first remark that, if a set A of Ω consists of a single point a of Ω, then A is nowhere dense in Ω. Consider any non-empty open set G of Ω. If $a \notin G$, the non-empty open subset G of G does not meet A. If $a \in G$, as a is not an isolated point of Ω, there is a point b of G other than a. Then, for all sufficiently small positive ϵ, the sphere $S(b, \epsilon)$, of points x with $\rho(x, b) < \epsilon$, is a non-empty open subset of G that does not meet A. Thus A is nowhere dense in Ω.

Now consider the class $\mathscr{R}$ of all open sets containing the countable set D. We show that, if $R \in \mathscr{R}$, then $\Omega \backslash R$ is nowhere dense in Ω. Suppose $R \in \mathscr{R}$ and G is a non-empty open set of Ω. Then, for some integer i, we have

$$d_i \in V_i \subset G.$$

Then

$$d_i \in D \subset R.$$

So $G \cap R$ is a non-empty open subset of G that does not meet $\Omega \backslash R$. Thus $\Omega \backslash R$ is nowhere dense in Ω.

Each open set G of Ω is of the form

$$G = \bigcup_{V_i \subset G} V_i,$$

and so is of the form

$$G = \bigcup_j V_{i(j)},$$

where $i(1), i(2), \ldots$ is the finite or infinite sequence of those integers i with $V_i \subset G$. Thus the open set G is determined by the sequence $i(1), i(2), \ldots$. Hence the cardinal of the class $\mathscr{G}$ of open sets, and *a fortiori* the cardinal of the class $\mathscr{R}$, is at most the cardinal $\mathbf{c}$ of the continuum.

Let ω be the least ordinal with cardinal equal to that of $\mathscr{R}$. Then the sets of $\mathscr{R}$ can be arranged in a transfinite sequence

$$R_\tau \quad (\tau < \omega).$$

As we are assuming the continuum hypothesis, each ordinal τ less than ω has cardinal less than $\mathbf{c}$, and so has a finite cardinal or the cardinal $\aleph_0$ of the integers. Thus, for each $\tau < \omega$, the number of sets R_σ, with $\sigma < \tau$, is countable.

We now choose a transfinite sequence

$$x_\tau \quad (\tau < \omega),$$

of points. We take x_1 to be any point of R_1. Now suppose that τ is any ordinal less than ω and that x_σ has been chosen for all ordinals σ less than τ in a way that ensures that:

(*a*) $x_\sigma \in \bigcap_{\kappa \leqslant \sigma} R_\kappa \quad (\text{for } \sigma < \tau),$

(*b*) $x_\kappa \neq x_\sigma \quad (\text{for } \kappa < \sigma < \tau).$

We proceed to describe how an appropriate point x_τ can be chosen. Consider the set

$$E_\tau = \Big(\bigcup_{\sigma<\tau} \{x_\sigma\}\Big) \cup \Big(\bigcup_{\sigma \leqslant \tau} (\Omega \backslash R_\sigma)\Big)$$

of those points we need to avoid when choosing x_τ. As we have shown that each one point set and each set $\Omega \backslash R$ with $R \in \mathscr{R}$ is a nowhere dense set, and as τ being less than ω is a countable ordinal, it follows that E_τ is a countable union of nowhere dense sets. Hence, by the Baire category theorem (see the digression at the end of this proof), E_τ does not exhaust the whole space Ω and we can choose a point x_τ not in E_τ. With this choice

(*a*) $x_\tau \in \bigcap_{\kappa \leqslant \sigma} R_\kappa \quad (\text{for } \sigma \leqslant \tau),$

(*b*) $x_\kappa \neq x_\sigma \quad (\text{for } \kappa < \sigma \leqslant \tau).$

In this way the whole transfinite sequence

$$x_\tau \quad (\tau < \omega),$$

can be chosen so that

(a) $x_\tau \in \bigcap_{\sigma \leqslant \tau} R_\sigma$ (for $\tau < \omega$),

(b) $x_\sigma \neq x_\tau$ (for $\sigma < \tau < \omega$).

We now take
$$C = \bigcup_{\tau < \omega} \{x_\tau\},$$

and prove that C satisfies our requirements. To prove that C is uncountable, it suffices to prove that ω is uncountable, i.e. that the class $\mathscr{R}$ is uncountable. Now provided x is not in the countable set D, the set $\Omega \backslash \{x\}$ is in $\mathscr{R}$, and different points x not in D lead to different sets $\Omega \backslash \{x\}$ in $\mathscr{R}$. So it suffices to prove that $\Omega \backslash D$ is uncountable, or that Ω is uncountable. But, if Ω were countable, it would be a countable union of one point sets, and so a countable union of nowhere-dense sets, contrary to the Baire category theorem. Hence C is certainly uncountable.

Now suppose that R is any open set that contains D. Then $R = R_\sigma$, for some ordinal σ less than ω. Then, by our choice of the points x_τ with $\sigma \leqslant \tau < \omega$, we have

$$x_\tau \in R_\sigma \quad (\text{for } \sigma \leqslant \tau < \omega).$$

So the only points of C not in $R = R_\sigma$ belong to the countable set of points x_κ with $1 \leqslant \kappa < \sigma$. Thus C is concentrated on the countable set, D, as required.

Digression. We digress to outline the proof of the Baire category theorem that we have used in the above proof. This famous theorem asserts that a *non-empty complete metric space is of the second category*. Suppose Ω were a non-empty complete metric space that was a countable union

$$\Omega = \bigcup_{n=1}^{\infty} N_n,$$

of nowhere dense sets N_n, $n = 1, 2, \ldots$, one set perhaps being repeated infinitely often in the union. As N_1 is nowhere dense in Ω, there is a non-empty open set G_1 that does not meet N_1. So for some radius r_1 less than 1, and some point x_1, the closed sphere $\Sigma(x_1, r_1)$ of points x with $\rho(x, x_1) \leqslant r_1$ does not meet N_1. As the sphere $S(x_1, r_1)$ of points x with $\rho(x, x_1) < r_1$ is non-empty and open and N_2 is nowhere dense, there is a radius r_2 less than $\frac{1}{2}$, and a point x_2 so that the closed sphere $\Sigma(x_2, r_2)$ is contained in $\Sigma(x_1, r_1)$ but does not meet N_2. Proceeding inductively we construct a sequence of points $x_1, x_2, \ldots$, a sequence of positive numbers $r_1, r_2, \ldots$, tending to zero, defining a nested sequence

$$\Sigma(x_1, r_1), \quad \Sigma(x_2, r_2), \ldots$$

of closed spheres with

$$\Sigma(x_1, r_1) \supset \Sigma(x_2, r_2) \supset \ldots,$$

$$\Sigma(x_i, r_i) \cap N_i = \varnothing \quad (i = 1, 2, \ldots).$$

As Ω is complete there is a point x common to all the spheres

$$\Sigma(x_i, r_i) \quad (i = 1, 2, \ldots),$$

which thus lies in none of the sets N_i $(i = 1, 2, \ldots)$, and so does not belong to Ω. This contradiction proves the theorem.

Remark. As far I know no uncountable set concentrated on a countable set has been constructed without use of the continuum hypothesis.

Theorem 39. *Let C be an uncountable set, in a metric space Ω, that is concentrated on a countable set. Then $\mu^h(C) = 0$, for all h in $\mathscr{H}_0$.*

Proof. Suppose that C is concentrated on the countable set D. Let $d_1, d_2, \ldots$ be an enumeration of the points of D.

Let h be any function of $\mathscr{H}_0$ and let $\epsilon > 0$, $\delta > 0$ be given. As $h(t) \to 0$ as $t \to 0$, we can choose a sequence $r_1, r_2, \ldots$ of radii so that

$$2r_i < \delta \quad (i = 1, 2, \ldots),$$

$$\sum_{i=1}^{\infty} h(2r_i) < \tfrac{1}{2}\epsilon.$$

Then the sequence of open spheres

$$S(d_i, r_i) \quad (i = 1, 2, \ldots),$$

has a union covering D. As C is concentrated on D, the points of C not covered by the union can be enumerated as a sequence $c_1, c_2, \ldots$, which might terminate or be empty. Now C is covered by the system of spheres

$$S(d_1, r_1), S(d_2, r_2), \ldots, \quad S(c_1, r_1), S(c_2, r_2), \ldots,$$

all of diameter less than δ, with

$$h(S(d_1, r_1)) + h(S(d_2, r_2)) + \ldots + h(S(c_1, r_1)) + h(S(c_2, r_2)) + \ldots \leqslant 2\sum_{i=1}^{\infty} h(2r_i) < \epsilon.$$

So

$$\mu^h_\delta(C) < \epsilon.$$

As h in $\mathscr{H}_0$, $\epsilon > 0$ and $\delta > 0$ are arbitrary the result follows.

Proof of theorem 37. The result follows immediately by combination of theorems 38 and 39.

§4 Comparison theorems

We establish in this section a group of theorems comparing Hausdorff measures corresponding to different functions in $\mathscr{H}$. We introduce a partial order into the family $\mathscr{H}$, by writing

$$g \prec h,$$

and by saying that g corresponds to a smaller generalised dimension than h, if

$$h(t)/g(t) \to 0 \quad \text{as} \quad t \to 0.$$

Note that this partial order is not a total order as there are functions g, h in $\mathscr{H}$ for which the ratios $h(t)/g(t)$, $g(t)/h(t)$ both oscillate as t tends to zero; indeed there are such functions in $\mathscr{H}$ with

$$\liminf_{t\to 0} h(t)/g(t) = 0, \qquad \limsup_{t\to 0} h(t)/g(t) = +\infty.$$

Although they would not lead to the same measures, functions g, h with the property

$$0 < \liminf_{t\to 0} h(t)/g(t) \leqslant \limsup_{t\to 0} h(t)/g(t) < +\infty,$$

will be regarded as essentially equivalent.

4.1. To state our results we need

***Definition* 27.** *If a measure is defined on a space Ω, a set in Ω is said to be of σ-finite measure, if it can be expressed in some way as a countable union of sets of finite measure; otherwise it is said to be of non-σ-finite measure.*

We first prove that, if g, h are functions of $\mathscr{H}$ with $g \prec h$, the sets E that are reasonably small when regarded from the point of view of μ^g-measure, i.e. the sets of σ-finite μ^g-measure, are very small (i.e. have zero μ^h-measure) when regarded from the point of view of μ^h-measure. We then give examples that show that the class of Hausdorff measures, taken as a whole, do *not* provide us with an unambiguous method of determining whether or not a set A is smaller, about the same size or larger than another set B; the set that seems the larger may depend on the particular Hausdorff measure we choose to use in making the comparison. A further result is that, if a set E has zero μ^h-measure for some h in $\mathscr{H}$, then there is a sense in which E must be a very small set; there will be a g in $\mathscr{H}$ with $g \prec h$ for which the μ^g-measure of E is still zero. We mention, but do not prove, a corresponding dual result for some sets of non-σ-finite μ^h-measure.

The first result is straightforward.

***Theorem* 40.** *Let g, h be functions in $\mathscr{H}$ with $g \prec h$. If a set E in a metric space Ω has σ-finite μ^g-measure, it has zero μ^h-measure.*

Proof. First suppose that E is a set in Ω with finite μ^g-measure. Let $\mu^g(E) = M$. Let $\epsilon > 0$ and $\delta > 0$ be given. As

$$h(t)/g(t) \to 0 \quad \text{as} \quad t \to 0,$$

we can choose t_0, without $0 < t_0 < \delta$, so that

$$\frac{h(t)}{g(t)} < \frac{\epsilon}{M+1} \quad \text{for} \quad 0 < t \leqslant t_0.$$

Choose a covering $\{S_i\}$ of E satisfying

$$E \subset \bigcup S_i, \qquad d(S_i) \leqslant t_0, \qquad \Sigma g(S_i) < M+1.$$

Now the condition $g \prec h$ implies that $h(0) = 0$. So

$$h(t) \leqslant \frac{\epsilon}{M+1} g(t) \quad (\text{for} \quad 0 \leqslant t \leqslant t_0),$$

and
$$h(S_i) \leqslant \frac{\epsilon}{M+1} g(S_i) \quad (\text{for} \quad i = 1, 2, \ldots),$$

as $d(S_i) \leqslant t_0$ for $i = 1, 2, \ldots$. Hence

$$\sum_{i=1}^{\infty} h(S_i) \leqslant \frac{\epsilon}{M+1} \sum_{i=1}^{\infty} g(S_i) < \epsilon,$$

while
$$E \subset \bigcup S_i, \qquad d(S_i) \leqslant \delta.$$

Thus
$$\mu_\delta^h(E) < \epsilon.$$

As ϵ, δ are arbitrary positive numbers this implies that $\mu^h(E) = 0$.

Now, if E has σ-finite μ^g-measure, $E = \bigcup_{i=1}^{i=\infty} E_i$, for some sequence $\{E_i\}$ of sets of finite μ^g-measure. By the last paragraph, each set E_i is of zero μ^h-measure. Hence

$$\mu^h(E) \leqslant \sum_{i=1}^{\infty} \mu^h(E_i) = 0,$$

and $\mu^h(E) = 0$ as required.

Corollary. *Let f, g, h be functions in $\mathscr{H}$ with $f \prec g \prec h$. If a set E in a metric space Ω has σ-finite positive μ^g-measure, then E has zero μ^h-measure and non-σ-finite μ^f-measure.*

Proof. As E has σ-finite μ^g-measure, it follows from the theorem that E has zero μ^h-measure. If E had σ-finite μ^f-measure, it would follow from the theorem that E would have zero μ^g-measure; hence E must have non-σ-finite μ^f-measure.

4.2. The theory of Hausdorff measures would be more satisfactory if one could always compare two sets in the same space and say that one or other of the two must be the smaller from the point of view of Hausdorff measures. But this is by no means the case. The lack of a total ordering among the functions of $\mathscr{H}$ is reflected by a lack of a total ordering among sets. This was first shown by H. G. Eggleston (1950*b*, 1951). Eggleston worked in Euclidean space and needed to impose appropriate conditions on the non-comparable functions f, h of $\mathscr{H}$. Our result will be rather easier to prove as we will take the liberty to choose our space to suit our problem.

***Theorem* 41.** *Let f and h be functions in $\mathscr{H}_0$ with*

$$0 = \liminf_{t\to 0} h(t)/f(t) < \limsup_{t\to 0} h(t)/f(t) = +\infty. \tag{1}$$

Then there is a compact metric space Ω with compact subsets A and B with A of zero μ^h- measure and of non-σ-finite μ^f-measure and with B of non-σ-finite μ^h-measure and of zero μ^f-measure.

Proof. Write
$$g(t) = \sqrt{\{f(t)\,h(t)\}}.$$

It is easy to verify that g then belongs to $\mathscr{H}_0$. Using (1) we choose sequences $\{a_r\}$, $\{b_r\}$ of positive numbers decreasing to the limit zero and satisfying

$$\lim_{r\to\infty} h(a_r)/f(a_r) = 0, \tag{2}$$

$$\lim_{r\to\infty} f(b_r)/h(b_r) = 0. \tag{3}$$

We use the result and some of the details of the proof of theorem 36. In that proof the choice of the sequence $\{t_r\}$ was largely at our disposal; we had merely to ensure that it converged to zero so rapidly that the conditions (10) and (11) were satisfied. By replacing $\{a_r\}$ by a suitable subsequence of itself and taking this for our sequence $\{t_r\}$, we can satisfy these conditions with the function h of theorem 36 replaced by the function g of this theorem. Let Ω_A be the corresponding compact metric space with $\mu^g(\Omega_A) = 1$. Write $a_0 = +\infty$, and define new functions f^A, g^A, h^A of $\mathscr{H}_0$ by writing

$$f^A(t) = f(a_r), \quad g^A(t) = g(a_r), \quad h^A(t) = h(a_r),$$

for
$$a_r \leqslant t < a_{r-1} \quad (r = 1, 2, \ldots).$$

As the only possible positive diameters in Ω_A are the numbers of the sequence $\{a_r\}$ it is clear that the pairs f, f^A and g, g^A and h, h^A generate the same Hausdorff measures on Ω_A. In particular

$$\mu^{g^A}(\Omega_A) = 1.$$

But $$g^A(t)/f^A(t) = \sqrt{\{h^A(t)/f^A(t)\}} = \sqrt{\{h(a_r)/f(a_r)\}} \to 0,$$

as $t \to 0$, and

$$h^A(t)/g^A(t) = \sqrt{\{h^A(t)/f^A(t)\}} = \sqrt{\{h(a_r)/f(a_r)\}} \to 0,$$

as $t \to 0$. So $$f^A \prec g^A \prec h^A.$$

Consequently, by the corollary to theorem 39, Ω_A has zero μ^{h^A}-measure and non-σ-finite μ^{f^A}-measure. Hence Ω_A has zero μ^h-measure and non-σ-finite μ^f-measure.

Now, working similarly with the sequence $\{b_r\}$ we can construct a compact metric space Ω_B, which we can take to be a subset of the interval $[2, 3]$, with Ω_B of non-σ-finite μ^h-measure and of zero μ^f-measure. We make $\Omega_A \cup \Omega_B$ into a compact metric space, by extending the metric ρ defined separately on Ω_A and Ω_B, by taking

$$\rho(a, b) = d(\Omega_A) + d(\Omega_B)$$

for all pairs a, b with $a \in \Omega_A$ and $b \in \Omega_B$. Then writing

$$A = \Omega_A, \qquad B = \Omega_B$$

gives the required results.

4.3. Our next result is due to A. S. Besicovitch (1956).

***Theorem* 42.** *Let E be a set in a metric space Ω with $\mu^h(E) = 0$, for some h in $\mathscr{H}_0$. Then there is a function g in $\mathscr{H}_0$ with $g \prec h$ and $\mu^g(E) = 0$.*

Proof. We use the criterion established in theorem 32 for a set to have zero Hausdorff measure. As $\mu^h(E) = 0$, we can choose a sequence $\{S_i\}$ of sets so that

$$E \subset \bigcup_{i=j}^{\infty} S_i \quad (j = 1, 2, \ldots),$$

and $$\sum_{i=1}^{\infty} h(S_i) < +\infty.$$

We choose a strictly increasing sequence $J(1), J(2)$, of positive integers so large that

$$\sum_{i=J(j)}^{\infty} h(S_i) < 4^{-j} \quad (j = 1, 2, \ldots).$$

Choose a decreasing sequence of positive numbers $t_1, t_2, \ldots$ so that, for $j = 1, 2, \ldots, t_j$ is less than the least *non-zero* element of the set

$$\{d(S_1), d(S_2), \ldots, d(S_{J(j)})\}.$$

At the same time we choose the sequence $t_1, t_2, \ldots$ to tend to zero, and to satisfy the condition:

$$h(t_j) < \tfrac{1}{4}h(t_{j-1}) \quad (j = 2, 3, \ldots).$$

Then
$$h(t_j) \leqslant (\tfrac{1}{4})^{j-1} h(t_1) \quad (j = 1, 2, \ldots).$$

Now this choice ensures that all the sets S_i with $d(S_i) < t_j$, either have zero diameter, or have $i > J(j)$. Thus

$$\sum_{d(S_i)<t_j} h(S_i) \leqslant \sum_{i=J(j)+1}^{\infty} h(S_i) < 4^{-j} \quad (j = 1, 2, \ldots). \tag{4}$$

The function g, that we seek, can now be obtained (see p. 83) by taking:

$$g(0) = 0;$$
$$g(t) = \min\{2^j h(t), \quad 2^{j-1} h(t_j)\} \quad (\text{for } t_{j+1} \leqslant t < t_j);$$
$$g(t) = h(t) \quad (\text{for } t_1 \leqslant t).$$

It is easy to verify that the function g defined in this way is positive for $t > 0$, is monotonic increasing and continuous on the right for $t > 0$ (actually, it is continuous, if h is continuous). Further

$$g(t_{j+1}) = \min\{2^j h(t_{j+1}), \quad 2^{j-1} h(t_j)\} = 2^j h(t_{j+1}) < 2^j (\tfrac{1}{4})^j h(t_1),$$

so that $g(t_j) \to 0$ as $j \to \infty$. Hence $g(t) \to 0$ as $t \to 0$. Thus $g \in H_0$ as required.

For $t_{j+1} \leqslant t < t_j$, we have

$$g(t) = \min\{2^j h(t), \quad 2^{j-1} h(t_j)\} \geqslant \min\{2^j h(t), \quad 2^{j-1} h(t)\} = 2^{j-1} h(t).$$

Thus
$$h(t)/g(t) \to 0 \quad \text{as} \quad t \to 0,$$

and $g \prec h$.

Also
$$\sum_{i=1}^{\infty} g(S_i) = \sum_{d(S_i) \neq 0} g(S_i) \leqslant \sum_{d(S_i) \neq 0} 2^{j(i)} h(S_i),$$

where we define $j(i)$ so that

$$t_{j(i)+1} \leqslant d(S_i) < t_{j(i)},$$

with the convention $t_0 = +\infty$. So

$$\sum_{i=1}^{\infty} g(S_i) \leqslant \sum_{j=0}^{\infty} 2^j \{\sum_{t_{j+1} \leqslant d(S_i) < t_j} h(S_i)\}$$

$$\leqslant \sum_{t_1 \leqslant d(S_i)} h(S_i) + \sum_{j=1}^{\infty} 2^j \{\sum_{d(S_i)<t_j} h(S_i)\}$$

$$\leqslant \sum_{i=1}^{\infty} h(S_j) + \sum_{j=1}^{\infty} 2^j (\tfrac{1}{4})^j$$

$$= 1 + \sum_{i=1}^{\infty} h(S_i) < +\infty,$$

on using (4). Thus the criterion of theorem 32 is satisfied and $\mu^g(E) = 0$.

Remarks. It seems likely that this method of Besicovitch can be modified to show that if E has $\mu^{h_i}(E) = 0$, for a sequence $h_1, h_2, \ldots$ of functions of $\mathscr{H}_0$ with

$$h_{i+1} \prec h_i \quad (i = 1, 2, \ldots),$$

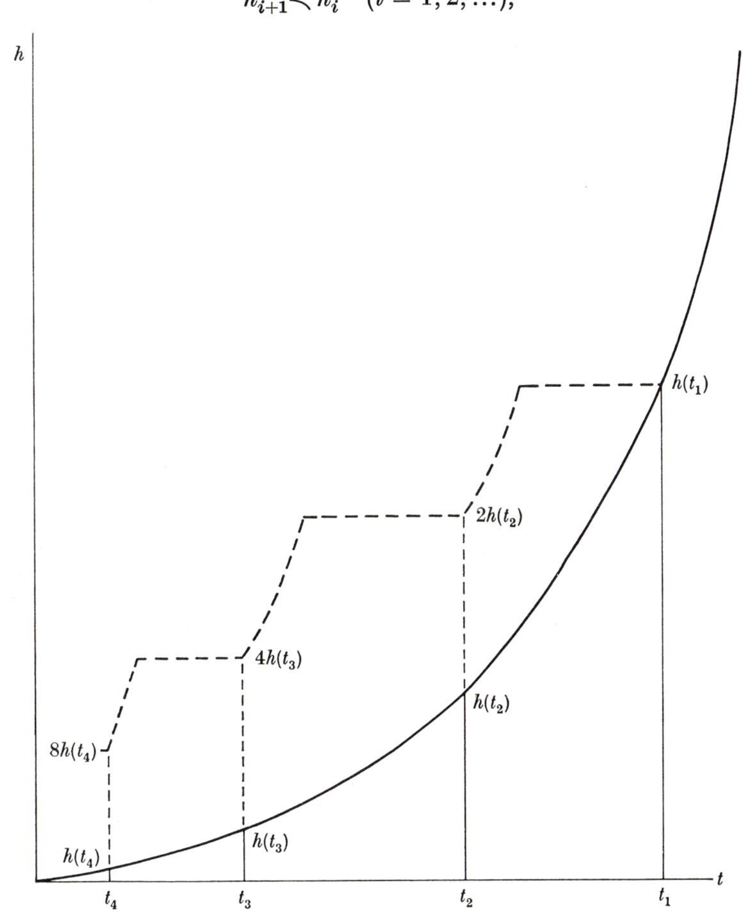

then there is a function g of $\mathscr{H}_0$ with

$$g \prec h_i \quad (i = 1, 2, \ldots),$$

and $\mu^g(E) = 0$. Indeed Rogers (1962) implies that this result was to have been included in C. A. Rogers and S. J. Taylor (1963), but they in fact could only prove this for σ-compact E.

4.4. It would be natural to suppose that, if a set E in a metric space Ω has non-σ-finite μ^g-measure for a function g of $\mathscr{H}_0$, then there is an h in $\mathscr{H}_0$ with $g \prec h$ for which E has non-σ-finite μ^h-measure. But analogy with theorem 37 suggests that this will not hold for all sets E. The result was proved by A. S. Besicovitch (1956b) for analytic sets in Euclidean space by methods that seemed to depend on the special properties of that space. It was proved by C. A. Rogers (1962) for continuous h and compact E, and M. Sion and D. Sjerve (1962) extended the result to Souslin-$\mathscr{F}$ subsets of a compact space. We remark that in theorems 35 and 40 we have essentially proved these results in the limiting case when g is taken to be the constant 1 so that μ^g reduces to 'counting measure'. But the existence of an uncountable set concentrated on a countable set (see theorem 39) shows that, given the continuum hypothesis, such results will not hold without some restriction on the class of sets considered (see A. S. Besicovitch, 1963c).

§5 Souslin sets

In preparation for the next two sections we prove some well-known results about Souslin sets. We have already introduced the Souslin operation in §7 of Chapter 1. The Souslin sets in a space are the sets obtained by application of the Souslin operation to closed sets. Our first aim will be to prove that each Borel set in a metric space is a Souslin set. We then show that, in a separable metric space, each Souslin set can be obtained by applying the Souslin operation to a system of closed sets that are nested in a suitable way and that have diameters tending to zero. Finally, we show that certain subsets of Souslin sets are compact.

5.1. We first prove the closure of the class of Souslin sets under the operations of countable unions and countable intersections. We remark that this class is, in fact, closed under the Souslin operation itself, but we shall not need to prove it.

***Theorem* 43.** *Let $\mathscr{A}$ be any class of sets containing Ω. Then the class of Souslin-$\mathscr{A}$ sets is closed under countable unions and under countable intersections.*

Proof. Let $\{A^{(m)}\}$ be any sequence of Souslin-$\mathscr{A}$ sets. Suppose that, for each m,

$$A^{(m)} = \bigcup_{\mathbf{i}\in\mathbf{I}} \bigcap_{n=1}^{\infty} A(m;\, \mathbf{i}|n),$$

where $A(m;\mathbf{i}|n) \in \mathscr{A}$ for all positive integers n and $\mathbf{i}$ in $\mathbf{I}$.

For each $\mathbf{j}$ in $\mathbf{I}$ and each integer n define the set $B(\mathbf{j}|n)$ by the equations

$$B(\mathbf{j}|n) = \Omega \quad \text{if} \quad n = 1,$$

$$B(\mathbf{j}|n) = A(j_1;\, j_2, j_3, \ldots, j_n) \quad \text{if} \quad n \geqslant 2.$$

Then
$$\bigcup_{\mathbf{j}\in\mathbf{I}}\bigcap_{n=1}^{\infty} B(\mathbf{j}|n) = \bigcup_{\mathbf{j}\in\mathbf{I}}\bigcap_{n=2}^{\infty} B(\mathbf{j}|n)$$
$$= \bigcup_{\mathbf{j}\in\mathbf{I}}\bigcap_{n=2}^{\infty} A(j_1; j_2, \ldots, j_n)$$
$$= \bigcup_{j_i=1}^{\infty}\bigcup_{\mathbf{i}\in\mathbf{I}}\bigcap_{n=1}^{\infty} A(j_1; i_1, \ldots, i_n)$$
$$= \bigcup_{m=1}^{\infty} A^{(m)},$$
showing that $\bigcup_{m=1}^{m=\infty} A^{(m)}$ is a Souslin-$\mathscr{A}$ set.

Now, each positive integer n has a unique representation in the form
$$n = 2^q l,$$
with $q = q(n)$, $l = l(n)$, q, l integral, and l odd.

For each $\mathbf{j}$ in $\mathbf{I}$ and each positive integer n, we introduce the set
$$C(\mathbf{j}|n) = A(q+1; j_{2^q}, j_{2^q.3}, \ldots, j_{2^q.l}), \tag{1}$$
the suffices of the 'j's being the odd multiples of 2^q not exceeding n. We show that
$$\bigcap_{m=1}^{\infty} A^{(m)} = \bigcup_{\mathbf{j}\in\mathbf{I}}\bigcap_{n=1}^{\infty} C(j|n). \tag{2}$$

If we choose a non-negative integer, q, we have, for $\mathbf{j}$ in $\mathbf{I}$,
$$\bigcap_{n=1}^{\infty} C(\mathbf{j}|n) \subset \bigcap_{l\ \text{odd}} C(\mathbf{j}|2^q l)$$
$$= \bigcap_{l\ \text{odd}} A(q+1; j_{2^q}, j_{2^q.3}, \ldots, j_{2^q.l})$$
$$= \bigcap_{n=1}^{\infty} A(q+1; \mathbf{i}_q|n),$$
where
$$\mathbf{i}_q = j_{2^q}, j_{2^q.3}, j_{2^q.5}, \ldots.$$
Hence
$$\bigcap_{n=1}^{\infty} C(\mathbf{j}|n) \subset \bigcup_{\mathbf{i}\in\mathbf{I}}\bigcap_{n=1}^{\infty} A(q+1; \mathbf{i}|n) = A^{(q+1)},$$
for all $q \geqslant 0$ and so
$$\bigcup_{\mathbf{j}\in\mathbf{I}}\bigcap_{n=1}^{\infty} C(\mathbf{j}|n) \subset \bigcap_{m=1}^{\infty} A^{(m)}. \tag{3}$$

On the other hand, if a is any point of $\bigcup_{m=1}^{m=\infty} A^{(m)}$, we can choose vectors
$$\mathbf{i}^{(m)} = i_1^{(m)}, i_2^{(m)}, \ldots,$$
in $\mathbf{I}$ with
$$a \in \bigcap_{n=1}^{\infty} A(m; \mathbf{i}^{(m)}|n),$$

for $m = 1, 2, \ldots$. We then form a vector $\mathbf{j}$, by taking

$$j_{2^q.l} = i^{(q+1)}_{(l+1)/2}.$$

Then for each integer $n = 2^q l$ we have

$$\begin{aligned} a \in A(q+1;\, i^{(q+1)}_1, i^{(q+1)}_2, \ldots, i^{(q+1)}_{(l+1)/2}) \\ = A(q+1;\, j_{2^q}, j_{2^q.3}, \ldots, j_{2^q.l}) \\ = C(\mathbf{j}|n). \end{aligned}$$

Thus
$$\bigcap_{m=1}^{\infty} A^{(m)} \subset \bigcup_{\mathbf{j} \in \mathbf{I}} \bigcap_{n=1}^{\infty} C(\mathbf{j}|n).$$

This together with (3) proves (2), and (2) together with (1) shows that $\bigcap_{m=1}^{m=\infty} A^{(m)}$ is a Souslin-$\mathscr{A}$ set.

***Theorem* 44.** *In a metric space Ω each Borel set is a Souslin-$\mathscr{F}$ set.*

Proof. Consider any closed set F of Ω. Then the set G_n of all points g with the property
$$\rho(g,f) < 1/n$$
for some point f of F, is an open set in Ω. Further

$$F = \bigcap_{n=1}^{\infty} G_n.$$

So each closed set in Ω is a $\mathscr{G}_\delta$-set. Taking complements, each open set in Ω is an $\mathscr{F}_\sigma$-set. Thus, by theorem 43, each open set of Ω has the property that both it and its complement are Souslin-$\mathscr{F}$ sets. Let $\mathscr{C}$ be the class of all Borel sets with this property (that they and their complements are Souslin-$\mathscr{F}$ sets). Then $\mathscr{C}$ contains the open sets. Further, by theorem 43, it follows immediately that $\mathscr{C}$ is closed under the operations of countable union and countable intersection. Further $\mathscr{C}$ is closed under complementation with respect to Ω. As the system of Borel sets is the minimal system of sets with these properties it follows that $\mathscr{C}$ must contain all Borel sets. So all Borel sets are Souslin-$\mathscr{F}$ sets.

***Theorem* 45.** *In a separable metric space each Souslin-$\mathscr{F}$ set A has a representation*

$$A = \bigcup_{\mathbf{i} \in \mathbf{I}} \bigcap_{n=1}^{\infty} F_{\mathbf{i}|n},$$

the sets $F_{\mathbf{i}|n}$ being closed sets with

$$d(F_{\mathbf{i}|n}) \leqslant 1/n \quad (\mathbf{i} \in \mathbf{I} \quad \text{and} \quad n = 1, 2, \ldots),$$

$$F_{\mathbf{i}|n+1} \subset F_{\mathbf{i}|n} \quad (\mathbf{i} \in \mathbf{I} \quad \text{and} \quad n = 1, 2, \ldots).$$

Proof. Let Ω be a separable metric space, with metric ρ. Let A be a Souslin-$\mathscr{F}$ set in Ω. Then A has a representation

$$A = \bigcup_{\mathbf{i} \in \mathbf{I}} \bigcap_{n=1}^{\infty} F^*_{\mathbf{i}|n}$$

the sets $F^*_{\mathbf{i}|n}$ being closed for all $\mathbf{i}$ in $\mathbf{I}$ and all $n \geqslant 1$.

As Ω is separable, there is a countable sequence $G_1, G_2, \ldots$ of non-empty open sets that is a base for the open sets of Ω. Let $g_1, g_2, \ldots$ be any sequence of points with g_i in G_i for $i = 1, 2, \ldots$. Then each non-empty open set G of Ω is a union of some of the sets $G_1, G_2, \ldots$, and so contains at least one of the points $g_1, g_2, \ldots$. Let $\Sigma(g_i, n)$ denote the closed sphere of points x with

$$\rho(x, g_i) \leqslant 1/(2n).$$

Then
$$d(\Sigma(g_i, n)) \leqslant 1/n, \tag{4}$$

for all positive integers i, n.

We note that, for fixed n, the sequence of sets $\Sigma(g_i, n), i = 1, 2, \ldots$ covers Ω. This follows, since, if x is any point of Ω, the sphere of points y with $\rho(x, y) < 1/(2n)$ is open and so contains a point, g_i say, of the sequence $\{g_i\}$, with the consequence that x lies in the corresponding set $\Sigma(g_i, n)$.

We now write

$$F_{\mathbf{i}|n} = \Big(\bigcap_{2k+1 \leqslant n} F^*_{i_1, i_3, \ldots, i_{2k+1}}\Big) \cap \Big(\bigcap_{2l \leqslant n} \Sigma(g_{i_{2l}}, 2l+1)\Big), \tag{5}$$

for all $\mathbf{i}$ in $\mathbf{I}$ and $n \geqslant 1$. Then, for all $\mathbf{i}$ in $\mathbf{I}$ and all $n \geqslant 1$, the set $F_{\mathbf{i}|n}$ is closed and has

$$d(F_{\mathbf{i}|n}) \leqslant d(\Sigma(g_{i_k}, k+1)),$$

where $k = 2[\tfrac{1}{2}n]$, so that

$$d(F_{\mathbf{i}|n}) \leqslant 1/(k+1) \leqslant 1/n,$$

by (4). Further, by (5), $\quad F_{\mathbf{i}|n+1} \subset F_{\mathbf{i}|n}$.

So it remains to verify that

$$A = \bigcup_{\mathbf{i} \in \mathbf{I}} \bigcap_{n=1}^{\infty} F_{\mathbf{i}|n}.$$

Clearly, by (5),
$$\bigcup_{\mathbf{i} \in \mathbf{I}} \bigcap_{n=1}^{\infty} F_{\mathbf{i}|n} \subset \bigcup_{\mathbf{i} \in \mathbf{I}} \bigcap_{n=1}^{\infty} F^*_{\mathbf{i}|n} = A.$$

But, if a is in A, we can choose a sequence of positive integers

$$i_1, i_3, \ldots, i_{2k+1}, \ldots$$

so that
$$a \in F^*_{i_1, i_3, \ldots, i_{2k+1}} \quad (k = 0, 1, 2, \ldots),$$

and we can choose a second sequence of positive integers $i_2, i_4, \ldots, i_{2l}, \ldots$ so that

$$a \in \Sigma(g_{i_{2l}}, 2l+1) \quad (l = 1, 2, \ldots).$$

Taking

$$\mathbf{i} = i_1, i_2, \ldots,$$

we see that

$$a \in F_{\mathbf{i}|n} \quad (n = 1, 2, \ldots).$$

So

$$a \in \bigcup_{\mathbf{i} \in \mathbf{I}} \bigcap_{n=1}^{\infty} F_{\mathbf{i}|n}.$$

Hence A has the required representation.

***Theorem* 46.** *Let Ω be a complete separable metric space. Suppose that a family $F_{\mathbf{i}|n}$ of closed sets satisfies the conditions*

$$d(F_{\mathbf{i}|n}) \leqslant 1/n \quad (\mathbf{i} \in \mathbf{I}, n = \mathbf{I}, n = 1, 2, \ldots),$$

$$F_{\mathbf{i}|n+1} \subset F_{\mathbf{i}|n} \quad (i \in \mathbf{I},\ n = 1, 2, \ldots).$$

Let $k_1, k_2, \ldots$ be an infinite sequence of positive integers, and let $\mathbf{K}, \mathbf{K}_m$ be the sets of $\mathbf{i} = i_1, i_2, \ldots$ in $\mathbf{I}$ with

$$1 \leqslant i_r \leqslant k_r \quad (r = 1, 2, \ldots),$$

or with

$$1 \leqslant i_r \leqslant k_r \quad (r = 1, 2, \ldots, m).$$

Then the set

$$C = \bigcup_{\mathbf{i} \in \mathbf{K}} \bigcap_{n=1}^{\infty} F_{\mathbf{i}|n}$$

is compact. Further, if G is any open set containing C, we have

$$\bigcup_{\mathbf{i} \in \mathbf{K}_m} \bigcap_{n=1}^{\infty} F_{\mathbf{i}|n} \subset G, \tag{6}$$

for m sufficiently large.

Proof. As Ω has a countable basis for its open sets, it will follow that C is compact, if we can show that any countable sequence of points of C has a subsequence converging to a point of C. Let $c_1, c_2, \ldots$ be any sequence of points of

$$C = \bigcup_{\mathbf{i} \in \mathbf{K}} \bigcap_{n=1}^{\infty} F_{\mathbf{i}|n}.$$

Then, for each positive integer r, we can choose a vector $\mathbf{i}(r)$ in $\mathbf{K}$ with

$$c_r \in F_{\mathbf{i}(r)|r}.$$

Now the first components $i_1(r), r = 1, 2, \ldots$ of the vectors

$$\mathbf{i}(r), \quad r = 1, 2, \ldots,$$

take only the values 1, 2, ..., k_1. So we can choose a subsequence R_1 of the natural numbers so that $i_1(r)$ takes a fixed value, say j_1, with $1 \leqslant j_1 \leqslant k$, for all r in R_1. By choosing a suitable subsequence R_2 of R_1 we can ensure that $i_2(r)$ takes a fixed value, say j_2, with

$$1 \leqslant j_2 \leqslant k_2,$$

for all r in R_2. Proceeding inductively in this way and then choosing a diagonal sequence R, we arrive at a sequence of integers $j_1, j_2, \ldots$ with

$$1 \leqslant j_s \leqslant k_s \quad (s = 1, 2, \ldots),$$

and with
$$i_s(r) = j_s,$$

for all r in R greater than or equal to the sth element r_s of R. Write $\mathbf{j} = j_1, j_2, \ldots$.

Then for all r, t in R not less than r_s, we have

$$i_\sigma(r) = i_\sigma(t) = j_\sigma \quad (\sigma = 1, 2, \ldots, s),$$

so that
$$\mathbf{i}(r)|s = \mathbf{i}(t)|s = \mathbf{j}|s,$$

and
$$c_r \in F_{\mathbf{i}(r)|s} = F_{\mathbf{j}|s}, \qquad c_t \in F_{\mathbf{i}(t)|s} = F_{\mathbf{j}|s}. \tag{7}$$

Hence
$$\rho(c_r, c_t) \leqslant d(F_{\mathbf{j}|s}) \leqslant 1/s.$$

Thus the subsequence $\{c_r\}_{r \in R}$ satisfies the Cauchy condition for convergence and converges to a point c say. As the set $F_{\mathbf{j}|s}$ is closed, for each s, it follows from (7) that

$$c \in F_{\mathbf{j}|s},$$

for each s. So the limit point c belongs to

$$\bigcap_{n=1}^{\infty} F_{\mathbf{j}|n}$$

and so to C, as $\mathbf{j} \in \mathbf{K}$. Thus C is compact.

Now let G be any open set containing C. Suppose that there are infinitely large m for which the inclusion (6) fails. Then we can choose an infinite sequence $c_1, c_2, \ldots$, of points with

$$c_m \in (\Omega \backslash G) \cap \bigcup_{\mathbf{i} \in \mathbf{K}_m} \bigcap_{n=1}^{\infty} F_{\mathbf{i}|n},$$

$m = 1, 2, \ldots$. Then we can choose vectors $\mathbf{i}(1), \mathbf{i}(2), \ldots$ with

$$c_m \in (\Omega \backslash G) \cap F_{\mathbf{i}(m)|m},$$
$$\mathbf{i}(m) \in \mathbf{K}_m.$$

We repeat the construction of the last two paragraphs and clearly arrive at a point c in C that is a limit point (and so a point) of the closed set $\Omega\backslash G$. This contradicts the hypothesis that $C \subset G$ and completes the proof of the theorem.

§6 The increasing sets lemma and its consequences

Since each Hausdorff measure μ^h with $h \in \mathscr{H}$, is regular, it follows (from theorem 9) that for any increasing sequence $\{E_n\}$ of sets

$$\mu^h\left(\bigcup_{n=1}^{\infty} E_n\right) = \sup_n \mu^h(E_n).$$

To obtain the main results of this section we need the corresponding result for μ^h_δ-measures, or a slightly weaker substitute. We prove this weaker result in the case when h is continuous and Ω is compact. This enables us to show that, if h is continuous, each Souslin-$\mathscr{F}$ set E, in a compact metric space, of positive μ^h-measure, contains compact subsets of positive μ^h-measure exceeding every number less than $\mu^h(E)$. The results of this section were obtained in a weak form by A. S. Besicovitch (1952), were proved in stronger forms by Roy O. Davies (1952) in Euclidean space and were later extended to and beyond the present level of generality by M. Sion and D. Sjerve (1962). More recently Roy O. Davies (1970) has made a more searching investigation of the circumstances in which an increasing sets lemma will hold.

6.1. We first prove

***Theorem* 47.** *Let Ω be a compact metric space. Let $\{E_n\}$ be any increasing sequence of sets. Suppose h is a continuous function in $\mathscr{H}_0$. Then, for all δ, η, with $0 < \delta < \eta$,*

$$\mu^h_\eta\left(\bigcup_{n=1}^{\infty} E_n\right) \leqslant \sup_n \mu^h_\delta(E_n) \leqslant \mu^h_\delta\left(\bigcup_{n=1}^{\infty} E_n\right). \tag{1}$$

Remark. Davies and Sion and Sjerve in fact prove the limiting case of this result in which $\delta = \eta$ so that equality holds in (1).

The proof will be based on a definition and on a form of Blaschke's 'Selection Theorem', see W. Blaschke (1916) or the proof of the lemma following the definition.

***Definition* 28.** *The distance $\rho(A, B)$ between two non-empty closed subsets of a metric space Ω is defined by*

$$\rho(A, B) = \max\{\epsilon(A, B), \epsilon(B, A)\},$$

0 0 0 0 0 0 0 0 0 0 0 0 0 0 0 0 0 0 0 0 0 0 0 0 0 0 0 0 0 0 0 0 0 0 0 0 0 0 0 0 0 0 0 0 0 0 0 0 0 0 0 0 0 0 0 0 0 0 0 0 0 0 0 0 0 0 0 0 0 0 0 0 0 0 0 0 0 0 0 0

1 2 3 4 5 6 7 8 9 10 11 12 13 14 15 16 17 18 19 20 21 22 23 24 25 26 27 28 29 30 31 32 33 34 35 36 37 38 39 40 41 42 43 44 45 46 47 48 49 50 51 52 53 54 55 56 57 58 59 60 61 62 63 64 65 66 67 68 69 70 71 72 73 74 75 76 77 78 79 80

1 1 1 1 1 1 1 1 1 1 1 1 1 1 1 1 1 1 1 1 1 1 1 1 1 1 1 1 1 1 1 1 1 1 1 1 1 1 1 1 1 1 1 1 1 1 1 1 1 1 1 1 1 1 1 1 1 1 1 1 1 1 1 1 1 1 1 1 1 1 1 1 1 1 1 1 1 1 1 1

2 2 2 2 2 2 2 2 2 2 2 2 2 2 2 2 2 2 2 2 2 2 2 2 2 2 2 2 2 2 2 2 2 2 2 2 2 2 2 2 2 2 2 2 2 2 2 2 2 2 2 2 2 2 2 2 2 2 2 2 2 2 2 2 2 2 2 2 2 2 2 2 2 2 2 2 2 2 2 2

3 3 3 3 3 3 3 3 3 3 3 3 3 3 3 3 3 3 3 3 3 3 3 3 3 3 3 3 3 3 3 3 3 3 3 3 3 3 3 3 3 3 3 3 3 3 3 3 3 3 3 3 3 3 3 3 3 3 3 3 3 3 3 3 3 3 3 3 3 3 3 3 3 3 3 3 3 3 3 3

4 4 4 4 4 4 4 4 4 4 4 4 4 4 4 4 4 4 4 4 4 4 4 4 4 4 4 4 4 4 4 4 4 4 4 4 4 4 4 4 4 4 4 4 4 4 4 4 4 4 4 4 4 4 4 4 4 4 4 4 4 4 4 4 4 4 4 4 4 4 4 4 4 4 4 4 4 4 4 4

5 5 5 5 5 5 5 5 5 5 5 5 5 5 5 5 5 5 5 5 5 5 5 5 5 5 5 5 5 5 5 5 5 5 5 5 5 5 5 5 5 5 5 5 5 5 5 5 5 5 5 5 5 5 5 5 5 5 5 5 5 5 5 5 5 5 5 5 5 5 5 5 5 5 5 5 5 5 5 5

6 6 6 6 6 6 6 6 6 6 6 6 6 6 6 6 6 6 6 6 6 6 6 6 6 6 6 6 6 6 6 6 6 6 6 6 6 6 6 6 6 6 6 6 6 6 6 6 6 6 6 6 6 6 6 6 6 6 6 6 6 6 6 6 6 6 6 6 6 6 6 6 6 6 6 6 6 6 6 6

7 7 7 7 7 7 7 7 7 7 7 7 7 7 7 7 7 7 7 7 7 7 7 7 7 7 7 7 7 7 7 7 7 7 7 7 7 7 7 7 7 7 7 7 7 7 7 7 7 7 7 7 7 7 7 7 7 7 7 7 7 7 7 7 7 7 7 7 7 7 7 7 7 7 7 7 7 7 7 7

8 8 8 8 8 8 8 8 8 8 8 8 8 8 8 8 8 8 8 8 8 8 8 8 8 8 8 8 8 8 8 8 8 8 8 8 8 8 8 8 8 8 8 8 8 8 8 8 8 8 8 8 8 8 8 8 8 8 8 8 8 8 8 8 8 8 8 8 8 8 8 8 8 8 8 8 8 8 8 8

9 9 9 9 9 9 9 9 9 9 9 9 9 9 9 9 9 9 9 9 9 9 9 9 9 9 9 9 9 9 9 9 9 9 9 9 9 9 9 9 9 9 9 9 9 9 9 9 9 9 9 9 9 9 9 9 9 9 9 9 9 9 9 9 9 9 9 9 9 9 9 9 9 9 9 9 9 9 9 9

Printed in U.S.A

where $\epsilon(C, D)$ is defined for all non-empty closed subsets C, D to be the infimum of the positive numbers ϵ with the property that each point d of D is within distance ϵ of some point of C.

Remark. It is very easy to verify that under this definition ρ becomes a metric on the space of non-empty closed subsets of Ω. This metric was much used by Hausdorff and will be called the Hausdorff metric. It is easy to verify that, if A, B are any two non-empty closed sets, their diameters satisfy

$$|d(A) - d(B)| \leqslant 2\rho(A, B).$$

So the diameter function is a uniformly continuous function on the space of non-empty closed sets under the metric ρ.

Lemma. (*Blaschke's Selection Theorem.*) *Let $\{F_n\}$ be a sequence of non-empty closed subsets of a compact metric space Ω, then there is a non-empty closed set F^* and a subsequence N of the natural numbers such that*

$$\rho(F_n, F^*) \to 0 \quad \textit{as} \quad n \to \infty \textit{ through } N.$$

Proof. Let $\mathscr{K}$ be the space of non-empty closed sets of Ω with the Hausdorff metric ρ. We have essentially to prove that $\mathscr{K}$ is compact under ρ, we start by proving that $\mathscr{K}$ is complete under ρ. Let $F_1, F_2, \ldots$ be any sequence of points of $\mathscr{K}$, i.e. any sequence of non-empty closed sets of Ω, and suppose that

$$\rho(F_n, F_m) \to 0 \quad \text{as} \quad n, m \to \infty.$$

We have to prove that there is a point F_0 of $\mathscr{K}$ such that

$$\rho(F_0, F_n) \to 0 \quad \text{as} \quad n \to \infty.$$

Let F_0 be the set of all points f of Ω with the property that f is a limit point of some sequence $f_1, f_2, \ldots$ with $f_i \in F_i$ for $i = 1, 2, \ldots$. As the sets $F_1, F_2, \ldots$ are non-empty, we can choose a sequence $f_1, f_2, \ldots$ with $f_i \in F_i$ for $i = 1, 2, \ldots$; as Ω is compact this sequence has at least one limit point. Hence F_0 is non-empty. When $f^{(0)}$ is the limit of a sequence of points $\{f^{(n)}\}$ of F_0 there are sequences $\{f_i^{(n)}\}_{i=1}^{i=\infty}$, $n = 1, 2, \ldots$ converging to the points $f^{(n)}$, $n = 1, 2, \ldots$ and with

$$f_i^{(n)} \in F_i \quad (i, n = 1, 2, \ldots).$$

For each n we can choose an integer $i(n)$ with

$$\rho(f_{i(n)}^{(n)}, f^{(n)}) < 1/n,$$

and we can arrange that $i(1) < i(2) < \ldots$. Write

$$f_{i(n)}^{(0)} = f_{i(n)}^{(n)} \quad (\text{for } n = 1, 2, \ldots)$$

and $$f_i^{(0)} = f_i^{(1)} \quad (\text{if } i \text{ is not one of } i(1), i(2), \ldots).$$

Then $f_i^{(0)} \in F_i$, for $i = 1, 2, \ldots$, and $\{f_i^{(0)}\}$ has a subsequence $\{f_{i(n)}^{(0)}\}$ converging to $f^{(0)}$. Thus $f^{(0)} \in F_0$. So F_0 is closed.

Let $\epsilon > 0$ be given. Choose p so large that

$$\rho(F_n, F_m) < \tfrac{1}{2}\epsilon \quad (\text{for } n, m \geqslant p). \tag{2}$$

Consider any point f_0 of F_0. Then for some $m \geqslant p$ and f_m of F_m we have $\rho(f_m, f_0) < \frac{1}{2}\epsilon$. By (2), for each $n \geqslant p$ there will be a point f_n of F_n with $\rho(f_n, f_m) < \frac{1}{2}\epsilon$. Thus $f_n \in F_n$ and $\rho(f_n, f_0) < \epsilon$. As this holds for all f_0 of F_0, it follows that

$$\epsilon(F_n, F_0) \leqslant \epsilon, \tag{3}$$

for $n \geqslant p$. On the other hand, if $n \geqslant p$ and $f_n \in F_n$, then by (2) there is a sequence $\{g_i\}$ of points with

$$g_i \in F_i \quad (i = 1, 2, \ldots),$$

$$\rho(g_i, f_n) \leqslant \tfrac{1}{2}\epsilon \quad (i \geqslant p).$$

As Ω is compact $\{g_i\}$ has some limit point g. Then $g \in F_0$ and

$$\rho(g, f_n) \leqslant \tfrac{1}{2}\epsilon.$$

Hence
$$\epsilon(F_0, F_n) \leqslant \tfrac{1}{2}\epsilon.$$

Combining this with (3), we have

$$\rho(F_n, F_0) \leqslant \epsilon \quad (\text{for } n \geqslant p).$$

Thus $\{F_n\}$ converges to F_0 as required to complete the proof that $\mathscr{K}$ is a complete space.

The next step is to show that, if $\delta > 0$, then $\mathscr{K}$ can be covered by a *finite* collection of sets of diameter less than δ. As Ω is compact, Ω is covered by some finite selection of open spheres of radius $\frac{1}{2}\delta$. Let $S(1), S(2), \ldots, S(n)$ be such a selection of open spheres of radius $\frac{1}{2}\delta$. Let $\mathscr{P}$ denote the finite system of all partitions $P = (R; S)$ of the integers $1, 2, \ldots, n$ into two disjoint sets R, S, the first R being non-empty. For each P in $\mathscr{P}$ let $\mathscr{K}(P)$ be the set of those non-empty closed sets of Ω that meet the spheres $S(r)$, $r \in R$ but do not meet the spheres $S(s)$, $s \in S$. We show that, if F and H are two sets of $\mathscr{K}(P)$ for some P in $\mathscr{P}$, we have $\rho(F, H) \leqslant \delta$. For, if $f \in F$, then f lies in one of the spheres $S(1), S(2), \ldots, S(n)$, as these cover Ω, but lies in none of the spheres $S(s)$ with $s \in S$, as these do not meet F; so f lies in one of the spheres $S(r)$ with $r \in R$. As each of these spheres $S(r), r \in R$ meets H, there will be a point h of H in the sphere $S(r)$, $r \in R$ containing f. So $\rho(h, f) \leqslant \delta$. Thus $\epsilon(H, F) \leqslant \delta$. Similarly $\epsilon(F, H) \leqslant \delta$.Hence $\rho(H, F) \leqslant \delta$. As F and H may be any sets of $\mathscr{K}(P)$, it follows that $d(\mathscr{K}(\mathscr{P})) \leqslant \delta$. We show that the sets $\mathscr{K}(P)$, $P \in \mathscr{P}$ cover $\mathscr{K}$. For, if F is any non-empty closed set of Ω, we can take R to be the set of integers r with $1 \leqslant r \leqslant n$ and $F \cap S(r) \neq \varnothing$, and S to be the complementary set of integers s with $1 \leqslant s \leqslant n$ and $F \cap S(r) = \varnothing$. Then R is non-empty and $P = (R; S)$ is a partition in $\mathscr{P}$ and $F \in \mathscr{K}(P)$. Thus the finite system $\mathscr{K}(P)$, $P \in \mathscr{P}$ satisfies our requirements.

We can now complete the proof of the compactness of $\mathscr{K}$ by general arguments operating in $\mathscr{K}$. Let $\{F_n\}$ be any sequence of points of $\mathscr{K}$. We make an inductive choice of infinite subsequences

$$N_0, N_1, N_2, \dots$$

of the natural numbers with

$$N_0 \subset N_1 \subset N_2 \subset \dots$$

Take $N_0 = \{1, 2, \dots, i, \dots\}$. When N_{n-1} has been chosen for some $n \geqslant 1$, we choose a finite collection of sets of $\mathscr{K}$ each of diameter less than $1/n$, the whole collection covering $\mathscr{K}$. Then at least one of these sets in $\mathscr{K}$ must contain an infinite number of the points of the sequence

$$F_i \quad (i \in N_{n-1}). \tag{4}$$

Let $\mathscr{K}_n$ denote one of these sets containing infinitely many of the points (4) and let N_n be the corresponding infinite subsequence of N_{n-1}. In this way we not only choose the sequence of subsequences $N_0, N_1, N_2, \dots$, but we also choose a sequence of sets $\mathscr{K}_1, \mathscr{K}_2, \dots$ of $\mathscr{K}$ with

$$d(\mathscr{K}_n) \leqslant 1/n \quad (n = 1, 2, \dots),$$

$$F_i \in \mathscr{K}_n \quad (\text{for } i \in N_n,\ n = 1, 2, \dots).$$

Thus we have
$$\rho(F_i, F_j) \leqslant d(\mathscr{K}_n) \leqslant 1/n,$$

if i, j belong to N_n. Now form a diagonal sequence N, taking the first element, $i(1)$ say, of N_1 as the first element of N. When the nth element $i(n)$ of N has been chosen, let $i(n+1)$ be the first element of the infinite sequence N_{n+1} that exceeds $i(n)$. Then N is a strictly increasing infinite sequence.

$$i(1), i(2), \dots,$$

and $i(r) \in N_s$ whenever $r \geqslant s$. Thus, if $r \geqslant s \geqslant n$, $i(r), i(s)$ belong to N_n so that $F_{i(r)}, F_{i(s)}$ belong to $\mathscr{K}_n$ and

$$\rho(F_{i(r)}, F_{i(s)}) \leqslant 1/n.$$

Thus the sequence $\{F_{i(r)}\}_{r=1}^{r=\infty}$ satisfies the Cauchy condition for convergence. As $\mathscr{K}$ is complete, there is a point F^* say, of $\mathscr{K}$ such that

$$\rho(F_n, F^*) \to 0 \quad \text{as} \quad n \to \infty \text{ through } N,$$

as required.

Proof of theorem 47. As Ω is compact it has at most a finite number of isolated points. Further, these points can contribute nothing to any μ_η^h- or μ_δ^h-measure of any set. So we may suppose that Ω has no isolated points.

Write

$$E = \bigcup_{n=1}^{\infty} E_n.$$

As $E_n \subset E$, for each n, we have

$$\sup_n \mu_\delta^h(E_n) \leqslant \mu_\delta^h(E) = \mu_\delta^h\left(\bigcup_{n=1}^{\infty} E_n\right).$$

So we have to prove that

$$\mu_\eta^h(E) \leqslant \sup_n \mu_\delta^h(E_n),$$

and so we may suppose that the supremum has a finite value, say

$$\sup_n \mu_\delta^h(E_n) = \lambda < +\infty.$$

Let $\epsilon > 0$ be given. Then, for each positive integer n, we may choose a sequence $\{F_i^{(n)}\}_{i=1}^{i=\infty}$ of closed sets with

$$E_n \subset \bigcup_{i=1}^{\infty} F_i^{(n)}, \qquad d(F_i^{(n)}) \leqslant \delta, \qquad \sum_{i=1}^{\infty} h(F_i^{(n)}) > \lambda + \epsilon.$$

If some of the sets $F_i^{(n)}$ have zero diameter (which happens when $F_i^{(n)}$ is empty or consists of a single point) we replace them by larger sets of positive diameter less than δ, without allowing the sum $\Sigma h(F_i^{(n)})$ to exceed $\lambda + \epsilon$; this is possible as h is continuous and no point of Ω is isolated.

For each n we suppose the sets re-ordered and renamed so that

$$\delta \geqslant d(F_1^{(n)}) \geqslant d(F_2^{(n)}) \geqslant \dots \geqslant d(F_i^{(n)}) \geqslant \dots. \tag{5}$$

This process is always possible, as the condition

$$\sum_{i=1}^{\infty} h(F_i^{(n)}) \leqslant \lambda + \epsilon < +\infty, \tag{6}$$

ensures that only a finite number of the sets $F_i^{(n)}$ can have diameter exceeding any given value.

By Blaschke's Selection Theorem we can choose a non-empty closed set F_1^* and a subsequence N_1 of the positive integers so that

$$\rho(F_1^{(n)}, F_1^*) \to 0 \quad \text{as} \quad n \to \infty \text{ through } N_1.$$

Then we can choose a non-empty closed set F^* and a subsequence N_2 of N_1 so that

$$\rho(F_2^{(n)}, F_2^*) \to 0 \quad \text{as} \quad n \to \infty \text{ through } N_2.$$

Proceeding inductively in this way and then choosing a diagonal sequence, we obtain an infinite sequence $\{F_i^*\}$ of non-empty closed sets and a sequence N of positive integers with the property that, for each i,

$$\rho(F_i^{(n)}, F_i^*) \to 0 \quad \text{as} \quad n \to \infty \text{ through } N. \tag{7}$$

Write

$$d_i = d(F_i^*) \quad (i = 1, 2, \ldots).$$

By (5) and the convergence (7), we have

$$\delta \geqslant d_1 \geqslant d_2 \geqslant \ldots \geqslant d_i \geqslant \ldots.$$

By (6), the convergence (7) and the continuity of h, we have

$$\sum_{i=1}^{\infty} h(d_i) \leqslant \lambda + \epsilon < +\infty.$$

Write

$$\sum_{i=1}^{\infty} h(d_i) = l.$$

As l is finite

$$d_i \to 0 \quad \text{as} \quad i \to \infty. \tag{8}$$

As h is continuous, for each i, we choose δ_i, with $d_i < \delta_i < \eta$, so close to d_i that

$$h(\delta_i) < h(d_i) + \epsilon . 2^{-i}.$$

Let V_i be the closed set of all points v of Ω with

$$\rho(v, f) \leqslant \tfrac{1}{2}(\delta_i - d_i),$$

for some point f of F_i^*. Then

$$d(V_i) \leqslant d_i + 2[\tfrac{1}{2}(\delta_1 - d_i)] = \delta_i,$$

so that

$$h(V_i) \leqslant h(\delta_i) < h(d_i) + \epsilon . 2^{-i},$$

and

$$\sum_{i=1}^{\infty} h(V_i) < \sum_{i=1}^{\infty} h(d_i) + \epsilon = l + \epsilon.$$

Further the convergence (7) ensures that, for each i,

$$F_i^{(n)} \subset V_i \quad \text{(for all sufficiently large } n \text{ in } N). \tag{9}$$

We write

$$V = \bigcup_{i=1}^{\infty} V_i$$

and study the sets $E_n \backslash V$, proving that for each n

$$\mu^h(E_n \backslash V) \leqslant \lambda - l + 3\epsilon.$$

Let n be given. Let $\delta^* > 0$ given. By (8) we can choose i^* so that

$$d_{i^*} < \delta^*,$$

and

$$\sum_{i=i^*+1}^{\infty} h(d_i) < \epsilon.$$

Then we can choose n^* in N with $n < n^*$ so that

$$F_i^{(n^*)} \subset V_i \quad (i = 1, 2, \ldots, i^*),$$

$$h(F_i^{(n^*)}) \geqslant h(d_i) - \epsilon . 2^{-i} \quad (i = 1, 2, \ldots, i^*),$$

$$d(F_{i^*}^{(n^*)}) < \delta^*.$$

These conditions ensure that

$$E_n \subset E_{n^*} \subset \bigcup_{i=1}^{\infty} F_i^{(n^*)},$$

$$\bigcup_{i=1}^{i^*} F_i^{(n^*)} \subset \bigcup_{i=1}^{\infty} V_i = V,$$

so that
$$E_n \backslash V \subset \bigcup_{i=i^*+1}^{\infty} F_i^{(n^*)}.$$

Further, for $i \geqslant i^* + 1$,

$$d(F_i^{(n^*)}) \leqslant d(F_{i^*}^{(n^*)}) < \delta^*.$$

Finally
$$\sum_{i=i^*+1}^{\infty} h(F_i^{(n^*)}) = \left\{\sum_{i=1}^{\infty} h(F_i^{(n^*)})\right\} - \left\{\sum_{i=1}^{i^*} h(F_i^{(n^*)})\right\}$$

$$\leqslant \lambda + \epsilon - \sum_{i=1}^{i^*} [h(d_i) - \epsilon . 2^{-i}]$$

$$< \lambda + 2\epsilon - \sum_{i=1}^{i^*} h(d_i)$$

$$< \lambda - l + 3\epsilon.$$

Hence
$$\mu_{\delta^*}^h(E_n \backslash V) \leqslant \lambda - l + 3\epsilon.$$

As δ^* may be arbitrarily small we deduce that

$$\mu^h(E_n \backslash V) \leqslant \lambda - l + 3\epsilon,$$

for all n. Thus
$$\mu^h(E \backslash V) = \mu^h\left(\bigcup_{n=1}^{\infty} \{E_n \backslash V\}\right)$$

$$= \sup_n \mu^h(E_n \backslash V)$$

$$\leqslant \lambda - l + 3\epsilon.$$

As V is covered by the sequence $\{V_i\}$ of sets of diameter less than η we deduce that

$$\mu_\eta^h(V) \leqslant \sum_{i=1}^{\infty} h(V_i) < l + \epsilon.$$

So
$$\mu_\eta^h(E) \leqslant \mu_\eta^h(V) + \mu_\eta^h(E \backslash V) \leqslant l + \epsilon + \lambda - l + 3\epsilon = \lambda + 4\epsilon.$$

As ϵ was an arbitrary positive number, it follows that

$$\mu_\eta^h(E) < \lambda,$$

so that
$$\mu_\eta^h\left(\bigcup_{n=1}^{\infty} E_n\right) \leqslant \sup_n \mu_\delta^h(E_n).$$

Thus (1) follows.

***Definition* 29.** *We will say that the increasing sets lemma holds for μ_δ^h-measure if, for each increasing sequence of sets $\{E_n\}$ and, for each $\eta > \delta$, we have*

$$\mu_\eta^h\left(\bigcup_{n=1}^{\infty} E_n\right) \leqslant \sup_n \mu_\delta^h(E_n) \leqslant \mu_\delta^h\left(\bigcup_{n=1}^{\infty} E_n\right). \tag{10}$$

So theorem 47 asserts that the increasing sets lemma holds for μ_δ^h-measure in any compact metric space, provided h is a continuous function in $\mathscr{H}_0$ and $\delta > 0$.

6.2. We can now obtain a result concerning the approximation of a Souslin-$\mathscr{F}$ set from within by compact sets.

***Theorem* 48.** *Let Ω be a complete separable metric space and let h be a function of $\mathscr{H}_0$. Suppose that the increasing sets lemma holds for μ_δ^h-measure for all $\delta > 0$. Let E be a Souslin-$\mathscr{F}$ set in Ω with*

$$\mu^h(E) > \lambda > 0,$$

for some real λ. Then E has a compact subset C with

$$\mu^h(C) > \lambda.$$

Proof. First choose ν with

$$\mu^h(E) > \nu > \lambda. \tag{11}$$

Then choose $\delta > 0$ so that $\mu_\delta^h(E) > \nu$.

Write
$$\delta(r) = \tfrac{1}{3}(1 + r^{-1})\,\delta, \tag{12}$$
for $r = 1, 2, \ldots$.

As E is a Souslin-$\mathscr{F}$ set it has a representation

$$E = \bigcup_{\mathbf{i} \in \mathbf{I}} \bigcap_{n=1}^{\infty} F_{\mathbf{i}|n},$$

the sets $F_{\mathbf{i}|n}$ being closed. As Ω is a separable metric space, it follows from theorem 45, that the representation may be chosen so that

$$d(F_{\mathbf{i}|n}) \leqslant 1/n \quad (\mathbf{i} \in \mathbf{I} \quad \text{and} \quad n = 1, 2, \ldots).$$

We suppose this done.

For each $\mathbf{i}$ in $\mathbf{I}$ and each n, we write

$$E(\mathbf{i}|n) = \bigcup_{\substack{\mathbf{j} \in \mathbf{I} \\ \mathbf{j}|n = \mathbf{i}|n}} \bigcap_{m=1}^{\infty} F_{\mathbf{j}|m}.$$

Then, clearly
$$E = \bigcup_{i_1=1}^{\infty} E(i_1).$$

As
$$\mu_\delta^h(E) > \delta, \qquad 0 < \delta(1) < \delta,$$

the increasing sets lemma, for $\mu_{\delta(1)}^h$-measure, enables us to choose a positive integer k_1 with

$$\mu_{\delta(1)}^h\left(\bigcup_{i_1=1}^{k_1} E(i_1)\right) > \nu.$$

Now
$$\bigcup_{i_1=1}^{k_1} E(i_1) = \bigcup_{i_2=1}^{\infty}\left\{\bigcup_{i_1=1}^{k_1} E(i_1, i_2)\right\}.$$

So we can repeat the argument. Proceeding inductively in this way we can choose a sequence of positive integers $k_1, k_2, \ldots$ so that

$$\mu_{\delta(r)}^h\Big(\bigcup_{i_1 \leqslant k_1, \ldots, i_r \leqslant k_r} E(i_1, i_2, \ldots, i_r)\Big) > \nu, \tag{13}$$

for $r = 1, 2, \ldots$.

We take C to be the set

$$C = \bigcup_{\mathbf{i} \in \mathbf{K}} \bigcap_{n=1}^{\infty} F_{\mathbf{i}|n},$$

where $\mathbf{K}$ is the set of all sequences $\mathbf{i} = i_1, i_2, \ldots$ with $0 \leqslant i_r \leqslant k_r$, for $r = 1, 2, \ldots$. By theorem 46, this set is compact.

Suppose we had
$$\mu^h(C) \leqslant \lambda.$$

Then
$$\mu_{\delta/3}^h(C) < \nu.$$

So we can find a sequence $\{G_i\}$ of open sets with

$$C \subset \bigcup_{i=1}^{\infty} G_i, \qquad d(G_i) \leqslant \tfrac{1}{3}\delta, \qquad \sum_{i=1}^{\infty} h(G_i) < \nu.$$

By the last part of theorem 46 we can choose m so large that

$$\bigcup_{\mathbf{i} \in \mathbf{K}_m} \bigcap_{n=1}^{\infty} F_{\mathbf{i}|n} \subset \bigcup_{i=1}^{\infty} G_i,$$

$\mathbf{K}_m$ being the set of all sequences $\mathbf{i} = i_1, i_2, \ldots$ with

$$1 \leqslant i_r \leqslant k_r \quad \text{for} \quad 1 \leqslant r \leqslant m.$$

Hence
$$\mu^h_{\delta(m)}\left(\bigcup_{\mathbf{i}\in\mathbf{K}_m}\bigcap_{n=1}^{\infty}F_{\mathbf{i}|n}\right) \leqslant \mu^h_{\delta/3}\left(\bigcup_{\mathbf{i}\in\mathbf{K}_m}\bigcap_{n=1}^{\infty}F_{\mathbf{i}|n}\right)$$
$$\leqslant \mu^h_{\delta/3}\left(\bigcup_{i=1}^{\infty}G_i\right)$$
$$\leqslant \sum_{i=1}^{\infty}h(G_i) < \nu.$$

As
$$\bigcup_{\mathbf{i}\in\mathbf{K}_m}\bigcap_{n=1}^{\infty}F_{\mathbf{i}|n} = \bigcup_{\substack{i_1\leqslant k_1,\ldots,i_m\leqslant k_m\\ \mathbf{i}\in\mathbf{I}}}\bigcap_{n=1}^{\infty}F_{\mathbf{i}|n} = \bigcup_{i_1\leqslant k_1,\ldots,i_m\leqslant k_m}E(i_1,i_2,\ldots,i_m),$$

we have a contradiction to (13). Hence we must have
$$\mu^h(C) \geqslant \mu^h_{\delta/3}(C) \geqslant \nu > \lambda,$$
as required.

Corollary 1. *Under the conditions of the theorem, E contains a σ-compact subset C with*
$$\mu^h(C) = \mu^h(E).$$

Proof. Immediate.

Corollary 2. *If Ω is a σ-compact metric space and h is a continuous function of $\mathscr{H}_0$, each Souslin-$\mathscr{F}$ set E in Ω contains a σ-compact subset C with*
$$\mu^h(C) = \mu^h(E).$$

Proof. As Ω is σ-compact we can choose an increasing sequence $\{K_n\}$ of compact sets with
$$\Omega = \bigcup_{n=1}^{\infty}K_n.$$

By theorem 9
$$\mu^h(E) = \mu^h\left(\bigcup_{n=1}^{\infty}\{E\cap K_n\}\right) = \sup_n \mu^h(E\cap K_n).$$

As the set E is Souslin-$\mathscr{F}$ and as K_n is closed, theorem 43 shows that $E\cap K_n$ is a Souslin-$\mathscr{F}$ set. Hence $E\cap K_n$ is a Souslin-$\mathscr{F}$ set when regarded as a subset of the compact metric space K_n. By theorems 47 and 48 applied to $E\cap K_n$ as a subset of K_n we can choose a σ-compact subset C_n of K_n with
$$\mu^h(C_n) = \mu^h(E\cap K_n).$$

Now each compact subset of K_n is compact in Ω. So C_n is σ-compact in Ω, and so is
$$C = \bigcup_{n=1}^{\infty}C_n.$$

Now
$$\mu^h(C) \geqslant \sup_n \mu^h(C_n) = \sup_n \mu^h(E\cap K_n) = \mu^h(E) \geqslant \mu^h(C).$$

So $\mu^h(C) = \mu^h(E)$ as required.

In these rather special circumstances, this second corollary extends to sets of infinite measure the result on approximation from within, proved in theorem 22 only for measurable sets of finite measure. In the case when E has finite measure, the equality

$$\mu^h(C) = \mu^h(E),$$

implies that

$$\mu^h(E\backslash C) = 0,$$

as C is μ^h-measurable. But we can no longer make this deduction when $\mu^h(E) = +\infty$. We ask whether, given a Souslin-$\mathscr{F}$ set E in a σ-compact metric space, it is always possible to choose a σ-compact set C contained in E with

$$\mu^h(E\backslash C) = 0.$$

This is always possible if E is of σ-finite μ^h-measure; but it is not possible in general. This impossibility was demonstrated by D. G. Larman (1965), who refined an earlier example of A. S. Besicovitch (1954) and constructed a $\mathscr{G}_\delta$-set E on the real line with the property that, if C is any $\mathscr{F}_\sigma$-set contained in E, the residual set $E\backslash C$ is always so large that the original set E can be covered by a countable union of translations of $E\backslash C$.

Returning to the consideration of theorem 48, one asks whether the compact set C can be taken to have non-σ-finite μ^h-measure in the case when E has non-σ-finite μ^h-measure. This was proved by R. O. Davies (1956) in Euclidean space, and extended to (and slightly beyond) the present level of generality by M. Sion and D. Sjerve (1962) (see p. 127).

The more difficult question of whether the compact set C of theorem 48 can always be chosen to have finite positive μ^h-measure will be discussed in the next section.

6.3. For a later application it is convenient to have a more abstract version of theorem 48. We introduce

***Definition* 30.** *A decreasing sequence of sets $E_1, E_2, \ldots$ with intersection E is said to converge strongly to E, if, for every open set G with $E \subset G$, we have $E_r \subset G$ for all sufficiently large r.*

Corollary 3 (*to theorem* 48). *Let μ be a measure defined in a complete separable metric space. Suppose that*

$$\sup_n \mu(E_n) = \mu\left(\bigcup_{n=1}^{\infty} E_n\right)$$

for every increasing sequence $\{E_n\}$ of sets. Suppose that

$$\mu(E_n) \to 0 \quad \textit{as} \quad n \to \infty,$$

whenever $\{E_n\}$ is a decreasing sequence of sets converging strongly to an intersection of zero μ-measure. Then each Souslin-$\mathscr{F}$ set of positive μ-measure has a compact subset of positive μ-measure.

Proof. This result follows without difficulty by the method used to prove theorem 48. Starting from the assumption that

$$E = \bigcup_{\mathbf{i} \in \mathbf{I}} \bigcap_{n=1}^{\infty} F_{\mathbf{i}|n}$$

is a Souslin-$\mathscr{F}$ set with $\quad \mu(E) > \nu > 0,$

the strong form of the increasing sets lemma enables us to choose $k_1, k_2, \ldots$ so that

$$\mu(\bigcup_{i_1 \leqslant k_1, \ldots, i_r \leqslant k_r} E(i_1, i_2, \ldots, i_r)) > \nu, \tag{14}$$

for $r = 1, 2, \ldots$. By theorem 46, and the proof of theorem 48, these sets

$$\bigcup_{i_1 \leqslant k_1, \ldots, i_r \leqslant k_r} E(i_1, i_2, \ldots, i_r) \quad (r = 1, 2, \ldots),$$

converge strongly to the compact subset C of E. By our last hypothesis and (14), it follows that C has positive μ-measure.

§7 The existence of comparable net measures and their properties

The results of this section, in as far as they go beyond results already in the literature, are joint work of Roy O. Davies and C. A. Rogers. The author is grateful to Dr Davies for allowing its first publication in this book.

We start with a definition of a net appropriate for our purposes.

***Definition* 31.** *A collection $\mathscr{N}$ of non-empty subsets of a metric space Ω will be called a net if it satisfies the conditions:*

(*a*) *if N_1 and N_2 belong to $\mathscr{N}$ then $N_1 \subset N_2$ or $N_2 \subset N_1$ or $N_1 \cap N_2 = \varnothing$;*

(*b*) *each point of Ω either belongs to a set of $\mathscr{N}$ of zero diameter or belongs to sets N of $\mathscr{N}$ having arbitrarily small diameters;*

(*c*) *$\mathscr{N}$ is countable;*

(*d*) *at most a finite number of sets of $\mathscr{N}$ contain any given set of $\mathscr{N}$;*

(*e*) *each set of $\mathscr{N}$ is an $\mathscr{F}_\sigma$-set.*

The net measures are the measures constructed by Method II from a pre-measure defined on $\mathscr{N} \cup \{\varnothing\}$ for some net $\mathscr{N}$. We shall also have occasion to study measures constructed by Method I from such pre-measures.

Such net measures were first introduced by A. S. Besicovitch (1952) in Euclidean space. They were constructed in some more general metric spaces by D. G. Larman (1967*b*, *c*). They are related to the measures of Hausdorff type discussed by E. V. Glivenko (1956) and later by Roy O. Davies (1969).

We shall first show that in Euclidean space a Hausdorff measure can be approximated reasonably well by an appropriate net measure. We mention an extension of this result to compact metric spaces that are not too large in a certain sense. We prove a corresponding result, for continuous h, when the space is a separable ultra-metric space. We then investigate the net measures and establish the strong form of the increasing sets lemma for their auxiliary measures (i.e. the Method I measures whose supremum defines the Method II net measure). This enables us to prove that compact sets of positive measure with respect to net measures satisfying a certain axiom, concerning decreasing sequences of compact sets, have compact subsets of finite positive measure. This is then extended, under suitable assumptions, to Souslin-$\mathscr{F}$ sets of positive measures. Finally the results on suitable specialization yield the results that, if A is a Souslin-$\mathscr{F}$ set of positive μ^h-measure, then A contains a compact subset of finite positive μ^h-measure, when h is continuous and the space Ω is either Euclidean or a complete separable ultra-metric space, or, if some results of Larman are taken for granted, for the spaces discussed by him. A final remark explains that the results do not hold without some restriction.

Net measures of the type discussed in this section are essential for the work of J. M. Marstrand (1954*a*) and D. G. Larman (1967*c*) on the properties of cartesian product sets.

7.1. We first give Besicovitch's construction in Euclidean space of such net-measures that are comparable to the Hausdorff measures.

***Theorem* 49.** *Let h be any function of $\mathscr{H}$. Then there is a net measure μ^τ in n-dimensional Euclidean space with*

$$\mu_\delta^h(E) \leqslant \mu_\delta^\tau(E) \leqslant 3^n 2^{n(n+1)} \mu_\delta^h(E), \tag{1}$$

for all δ with $0 < \delta < 1$ and all sets E. Further

$$\mu^h(E) \leqslant \mu^\tau(E) \leqslant 3^n 2^{n(n+1)} \mu^h(E), \tag{2}$$

for all sets E.

Proof. For each non-negative integer r and each set of n integers

$i_1, i_2, \ldots, i_n$ let $N_r(i_1, i_2, \ldots, i_n)$ be the set of all points $\mathbf{x} = (x_1, x_2, \ldots, x_n)$ with

$$2^{-r} i_j \leqslant x_j < 2^{-r}(i_j + 1) \quad (j = 1, 2, \ldots, n).$$

Take $\mathcal{N}$ to be the system of all these cubes

$$N_r(i_1, i_2, \ldots, i_n) \quad (r = 0, 1, \ldots, \quad i_1, \ldots, i_n = 0, \pm 1, \pm 2, \ldots).$$

The conditions (a) to (e) of definition 31 are clearly satisfied and $\mathcal{N}$ is a net. Indeed, this particular system of sets is the prototype net.

Let τ be the pre-measure on $\mathcal{N}$ obtained by writing

$$\tau(\varnothing) = 0, \quad \tau(N_r(i_1, i_2, \ldots, i_n)) = h(2^{-r}\sqrt{n}).$$

Let μ^τ be the net measure defined from the pre-measure τ on the net $\mathcal{N}$.

Let E be any set of Ω. Suppose that $0 < \delta < 1$. As the diameter of

$$N_r(i_1, i_2, \ldots, i_n) \quad \text{is} \quad 2^{-r}\sqrt{n},$$

it follows immediately that

$$\mu_\delta^h(E) \leqslant \mu_\delta^\tau(E).$$

But, if $\mu_\delta^h(E) < +\infty$, and $\epsilon > 0$ is given, we can choose a sequence $\{S_i\}$ of non-empty sets with

$$E \subset \bigcup_{i=1}^{\infty} S_i, \qquad d(S_i) \leqslant \delta, \qquad \sum_{i=1}^{\infty} h(S_i) < \mu_\delta^h(E) + \epsilon.$$

For each i choose the integer $r(i)$ for which

$$2^{-r(i)-1} \leqslant d(S_i) < 2^{-r(i)}.$$

Then S_i meets one of the cubes $N_{r(i)}(i_1, i_2, \ldots, i_n)$ and so is contained in the union of the 3^n cubes

$$N_{r(i)}(j_1, j_2, \ldots, j_n), \qquad j_k = i_k \quad \text{or} \quad i_k \pm 1 \quad \text{for} \quad k = 1, 2, \ldots, n.$$

So S_i is contained in a system $N_{i1}, N_{i2}, \ldots,$ of $3^n \,.\, 2^{n(n+1)}$ cubes of side $2^{-r(i)-n-1}$ and of diameter

$$d(N_{ij}) = 2^{-r(i)-n-1}\sqrt{n} < 2^{-r(i)-1} \leqslant d(S_i).$$

Hence

$$\mu_\delta^\tau(E) \leqslant \sum_{i,j} \tau(N_{ij}) \leqslant 3^n \,.\, 2^{n(n+1)} \sum_{i=1}^{\infty} h(S_i) \leqslant 3^n \,.\, 2^{n(n+1)} (\mu_\delta^h(E) + \epsilon).$$

Consequently $\qquad \mu_\delta^\tau(E) \leqslant 3^n \,.\, 2^{n(n+1)} \mu_\delta^h(E),$

as required. Finally (2) follows immediately from (1).

7.2. D. G. Larman (1967*b*, *c*) defines a finite-dimensional space to be a space Ω with zero (n)-measure for some positive integer n (here, as in Chapter 2, § 2, (n)-measure is μ^h-measure with $h(t) = t^n$). He proves that in any finite-dimensional compact metric space it is possible to construct a net $\mathcal{N}$ satisfying the conditions (a)–(e), and that given any h in $\mathcal{H}_0$ it is possible to define a pre-measure τ on $\mathcal{N}$ so that μ^τ is comparable to μ^h in that

$$\mu_\delta^h(E) \leqslant \mu_\delta^\tau(E) \leqslant N\mu_\delta^h(E), \tag{3}$$

for all $\delta > 0$ and for all E in Ω and some constant N. We do not give Larman's construction as it is rather complicated and may sooner or later be replaced by something simpler or more effective.

7.3. Another case when comparable net measures can be introduced is when the space is an ultra-metric space.

***Definition* 32.** *A metric space Ω with metric ρ is said to be an ultra-metric space, if, whenever x, y, z are three points of Ω,*

$$\rho(x,z) \leqslant \max\{\rho(x,y), \rho(y,z)\}.$$

***Theorem* 50.** *Let Ω be a separable ultra-metric space, and let h be any continuous function of $\mathcal{H}_0$. Then there is a net $\mathcal{N}$ on Ω and a pre-measure τ defined on $\mathcal{N}$ with*

$$\mu_\delta^h(E) \leqslant \mu_\delta^\tau(E) \leqslant 2\mu_{\delta/2}^h(E), \tag{4}$$

for all sufficiently small $\delta > 0$ and for all E of Ω. Further

$$\mu^h(E) \leqslant \mu^\tau(E) \leqslant 2\mu^h(E). \tag{5}$$

Proof. As h is continuous and non-decreasing we can choose a strictly decreasing sequence, $d_0, d_1, \ldots$, converging to 0, with

$$d_{i+1} > \tfrac{1}{2}d_i \quad (i = 0, 1, 2, \ldots), \tag{6}$$

and

$$h(d_{i+1}) > \tfrac{1}{2}h(d_i) \quad (i = 0, 1, 2, \ldots). \tag{7}$$

For each integer $i \geqslant 0$, and each point x of Ω we define $N(x, i)$ to be the set of all points y with $\rho(y, x) \leqslant d_i$. Now, if for some x, y in Ω and some integers i, j with $0 \leqslant i \leqslant j$ we have

$$N(x,i) \cap N(y,j) \neq \varnothing,$$

there is a point z with

$$\rho(x,z) \leqslant d_i, \qquad \rho(y,z) \leqslant d_j.$$

So, as Ω is ultra-metric,

$$\rho(x,y) \leqslant \max(d_i, d_j) = d_i.$$

Then, for any point u in $N(y,j)$ we have

$$\rho(x,y) \leqslant d_i, \qquad \rho(y,u) \leqslant d_j,$$

so that
$$\rho(x,u) \leqslant \max(d_i, d_j) = d_i,$$

and u is in $N(x,i)$. Thus the sets $N(x,i)$ satisfy the condition (a) for a net.

As the set $N(x,i)$ has diameter less than or equal to d_i, it is clear that the condition (b) is also satisfied.

As Ω is a separable metric space, we can choose a countable dense sequence $\{x_j\}$ of its points. But as Ω is ultra-metric, each set $N(x,i)$ is open. So, keeping i fixed, each of the different disjoint sets $N(x,i)$ must contain at least one point x_j. Thus the system of sets $N(x,i)$ is countable and condition (c) is satisfied.

For any x in Ω and any $i \geqslant 0$, the only sets $N(y,j)$ containing the set $N(x,i)$ must coincide with one of the finite system of sets

$$N(x,1),\ N(x,2),\ \ldots,\ N(x,k),$$

where $k+1$ is smallest integer for which $N(x,k+1)$ is a proper subset of $N(x,i)$ or if there is no such integer $k+1$, where k is the smallest integer for which
$$N(x,k) = \{x\}.$$

Thus the condition (d) is satisfied.

As the sets $N(x,i)$ are all closed (as well as open) the condition (e) is satisfied. This completes the proof that the system of sets $N(x,i)$ forms a net, $\mathcal{N}$ say.

We introduce a pre-measure τ on $\mathcal{N} \cup \{\varnothing\}$, taking

$$\tau(\varnothing) = 0,$$

$$\tau(N) = h(N), \qquad N \in \mathcal{N}.$$

It follows immediately that

$$\mu_\delta^h(E) \leqslant \mu_\delta^\tau(E),$$

for all $\delta > 0$ and for all E of Ω, as μ^τ is constructed by Method II from a restriction of the pre-measure used in the construction of μ^h.

Now consider any set E with $\mu_{\frac{1}{2}\delta}^h(E)$ finite for a given δ with $0 < \delta < d_0$. Let $\epsilon > 0$ be given. By (6) we can choose i^* so that

$$\tfrac{1}{2}\delta < d_{i^*} \leqslant \delta.$$

Then we can choose a cover $\{S_i\}$ of E with

$$E \subset \bigcup_{i=1}^{\infty} S_i, \qquad d(S_i) \leqslant \tfrac{1}{2}\delta, \qquad \sum_{i=1}^{\infty} h(S_i) \leqslant \mu_{\frac{1}{2}\delta}^h(E) + \epsilon.$$

We may suppose that each set S_i is non-empty and choose a point s_i in S_i for each i. For each integer i we either have $d(S_i) = 0$ or we can choose an integer $j(i)$ with $j(i) \geqslant i^*$ and

$$d_{j(i)+1} < d(S_i) \leqslant d_{j(i)}.$$

In the first case we can choose $j(i)$ so large that

$$j(i) > i^*, \qquad h(d_{j(i)}) < \epsilon . 2^{-i}.$$

Let $$N_i = N(s_i, j(i)) \quad (i = 1, 2, \ldots).$$

Then as S_i contains s_i and has diameter $d(S_i)$, not exceeding $d_{j(i)}$, we have $S_i \subset N_i\,(i = 1, 2, \ldots)$. So

$$E \subset \bigcup_{i=1}^{\infty} N_i.$$

Also $$d(N_i) \leqslant d_{j(i)} \leqslant d_{i^*} \leqslant \delta.$$

Further, using (7),

$$\begin{aligned}\sum_{i=1}^{\infty} \tau(N_i) &\leqslant \sum_{i=1}^{\infty} h(d_{j(i)}) \\ &\leqslant 2\sum_{i=1}^{\infty} h(d_{j(i)+1}) \\ &\leqslant 2\sum_{i=1}^{\infty} \{h(d(S_i)) + \epsilon . 2^{-i}\} \\ &\leqslant 2\epsilon + 2\sum_{i=1}^{\infty} h(S_i) \\ &\leqslant 2\mu^h_{\frac{1}{2}\delta}(E) + 4\epsilon.\end{aligned}$$

As $\epsilon > 0$ is arbitrary, this shows that

$$\mu^{\tau}_{\delta}(E) \leqslant 2\mu^h_{\frac{1}{2}\delta}(E),$$

for $0 < \delta < d_0$, as required. Finally (5) follows immediately from (4).

Corollary. *If Ω is compact, and ultra-metric there are at most a finite number of sets N of $\mathcal{N}$ with $d(N) > \delta$, for each $\delta > 0$.*

Proof. For each $i \geqslant 0$, the sets $N(x, i)$ with $x \in \Omega$ form an open cover of Ω, and so a finite subsystem covers Ω. The result follows.

7.4. By the results of Chapter 1 we easily obtain

Theorem 51. *Let μ be a net-measure constructed by Method II from a pre-measure τ defined on $\mathcal{N} \cup \{\varnothing\}$ where $\mathcal{N}$ is a net in a metric space Ω. Then μ is a regular metric measure.*

Proof. By theorems 16 and 19, μ is a metric measure and all Borel sets are measurable. By theorem 20, μ is $\mathcal{N}_{\sigma\delta}$-regular. As all sets of $\mathcal{N}$ are Borel sets, μ is regular.

7.5. We now obtain a strong form of the increasing sets lemma for the net measures constructed by Method I. H. Wegmann (1969*a*) obtains this result for Hausdorff measures in ultra-metric spaces.

***Theorem* 52.** *Let $\mathcal{N}$ be a net defined in a metric space Ω. Let ν be a measure constructed by Method I from a pre-measure τ defined on $\mathcal{N} \cup \{\varnothing\}$. If $\{E_n\}$ is any increasing sequence of sets*

$$\nu\left(\bigcup_{n=1}^{\infty} E_n\right) = \sup_n \nu(E_n).$$

Proof. Let $\{E^{(n)}\}$ be any increasing sequence of sets and write

$$E = \bigcup_{n=1}^{\infty} E^{(n)}.$$

As
$$\sup_n \nu(E^{(n)}) \leqslant \nu(E),$$

it suffices to suppose that

$$\sup_n \nu(E^{(n)}) = \lambda < +\infty,$$

and deduce that
$$\nu(E) \leqslant \lambda.$$

For any given $\epsilon > 0$, we can find for $n = 1, 2, \ldots$, a system $\{N_i^{(n)}\}_{i=1}^{i=\infty}$ of sets of $\mathcal{N} \cup \{\varnothing\}$ with

$$E^{(n)} \subset \bigcup_{i=1}^{\infty} N_i^{(n)}, \quad N_i^{(n)} \in \mathcal{N} \cup \{\varnothing\}, \quad \sum_{i=1}^{\infty} \tau(N_i^{(n)}) < \nu(E^{(n)}) + \epsilon . 2^{-n}.$$

Our aim is to select from the whole family of sets $\{N_i^{(n)}\}$ a subfamily that yields an efficient covering of E. As

$$E = \bigcup_{n=1}^{\infty} E^{(n)} \subset \bigcup_{n=1}^{\infty} \bigcup_{i=1}^{\infty} N_i^{(n)},$$

each point e of E lies in one of the sets $\{N_i^{(n)}\}$, say the set $N_{i(e)}^{(n(e))} \neq \varnothing$. As $\mathcal{N}$ is a net, $N_{i(e)}^{(n(e))}$ is included in only a finite number of sets of the system $\{N_i^{(n)}\}$, and so is included in some set of the system $\{N_i^{(n)}\}$ that is properly included in no other set of the system. Thus each point of E is included in one of the maximal sets of the system $\{N_i^{(n)}\}$. Let M_1, $M_2, \ldots$ be an enumeration of the maximal sets of the system $\{N_i^{(n)}\}$, augmented, if necessary by a terminal sequence $\varnothing, \varnothing, \ldots,$ of empty sets. Then

$$E \subset \bigcup_{j=1}^{\infty} M_j \quad (M_j \in \mathscr{N} \cup \{\varnothing\}). \tag{8}$$

We study the sum
$$\sum_{j=1}^{J} \tau(M_j),$$

for any fixed J. Each non-empty set M_j is one of the sets $\{N_i^{(n)}\}$. Let $n(j)$ be the smallest integer n such that M_j occurs in the sequence $\{N_i^{(n)}\}_{i=1}^{i=\infty}$. Write

$$m = \max_{1 \leqslant j \leqslant J} n(j).$$

For each n with $1 \leqslant n \leqslant m$, let $\mathscr{M}^{(n)}$ denote the family of those sets M_1, $M_2, \ldots, M_J$ that are non-empty and have $n(j) = n$. Then

$$\sum_{j=1}^{J} \tau(M_j) = \sum_{n=1}^{m} \sum_{M \in \mathscr{M}^{(n)}} \tau(M).$$

We compare the original covering of $E^{(n)}$ by the sets $\{N_i^{(n)}\}_{i=1}^{i=\infty}$ with certain new coverings constructed using the covering $\{N_i^{(m)}\}_{i=1}^{i=\infty}$ of $E^{(m)}$. In the first place we have

$$\sum_{M \in \mathscr{M}^{(n)}} \tau(M) + \sum_{\substack{i=1 \\ N_i^{(n)} \notin \mathscr{M}^{(n)}}}^{\infty} \tau(N_i^{(n)}) = \sum_{i=1}^{\infty} \tau(N_i^{(n)}) < \nu(E^{(n)}) + \epsilon . 2^{-n}. \tag{9}$$

Now let $\mathscr{L}^{(m,n)}$ denote those of the sets $N_i^{(m)}$ that meet one of the sets of the system $\mathscr{M}^{(n)}$. As all the sets in $\mathscr{M}^{(n)}$ are maximal sets in the system $\{N_i^{(l)}\}$, all such sets of $\mathscr{L}^{(m,n)}$ are contained in the corresponding set of $\mathscr{M}^{(n)}$. As $E^{(n)} \subset E^{(m)}$ we have

$$E^{(n)} \subset \bigcup_{i=1}^{\infty} N_i^{(m)},$$

and
$$E^{(n)} \cap \{\bigcup_{M \in \mathscr{M}^{(n)}} M\} \subset \bigcup_{L \in \mathscr{L}^{(m,n)}} L.$$

Thus the sets of $\mathscr{L}^{(m,n)}$ together with those sets $\{N_i^{(n)}\}_{i=1}^{i=\infty}$ not in $\mathscr{M}^{(n)}$ cover $E^{(n)}$. Hence

$$\sum_{L \in \mathscr{L}^{(m,n)}} \tau(L) + \sum_{\substack{i=1 \\ N_i^{(n)} \notin \mathscr{M}^{(n)}}}^{\infty} \tau(N_i^{(n)}) \geqslant \nu(E^{(n)}). \tag{10}$$

Subtracting (10) from (9) we deduce that

$$\sum_{M \in \mathscr{M}^{(n)}} \tau(M) < \sum_{L \in \mathscr{L}^{(m,n)}} \tau(L) + \epsilon . 2^{-n}.$$

Summing over n, we have

$$\sum_{j=1}^{J} \tau(M_j) = \sum_{n=1}^{m} \sum_{M \in \mathcal{M}^{(n)}} \tau(M)$$

$$< \sum_{n=1}^{m} \{ \sum_{L \in \mathcal{L}^{(m,n)}} \tau(L) + \epsilon . 2^{-n} \}$$

$$\leqslant \sum_{i=1}^{\infty} \tau(N_i^{(m)}) + \epsilon < \lambda + 2\epsilon.$$

Hence

$$\sum_{j=1}^{\infty} \tau(M_j) \leqslant \lambda + 2\epsilon. \tag{11}$$

Now by (8) and (11) we have

$$\nu(E) \leqslant \lambda + 2\epsilon.$$

As $\epsilon > 0$ is arbitrary this gives

$$\nu(E) \leqslant \lambda,$$

as required.

7.6. If $\mathcal{N}$ is a net and τ is a pre-measure defined on $\mathcal{N} \cup \{\varnothing\}$ the net measure μ^τ constructed by Method II from τ takes the form

$$\mu^\tau(E) = \sup_{\delta > 0} \nu_\delta^\tau(E), \tag{12}$$

where ν_δ^τ is the auxiliary measure constructed by Method I from the restriction τ_δ of τ to $\mathcal{N}_\delta \cup \{\varnothing\}$ where $\mathcal{N}_\delta$ is the subset of $\mathcal{N}$ of all sets of diameter not exceeding δ. So μ^τ can be regarded as a limit measure as the sets of largest diameter are removed from $\mathcal{N}$. For the sequel it will be convenient to have an alternative method of obtaining μ^τ as a limit measure as sets are removed *one-at-a-time* from $\mathcal{N}$.

***Theorem* 53.** *If $\mathcal{N}$ is a net in a metric space Ω, the sets of $\mathcal{N}$ of positive diameter can be ordered in sequence $N_1, N_2, \ldots$ so that:*
if

$$1 \leqslant r < s,$$

then either $\quad N_r \supset N_s \quad$ *and* $\quad N_r \neq N_s \quad$ *or* $\quad N_r \cap N_s = \varnothing.$

If $N_1, N_2, \ldots$ is such a sequence, and τ is a pre-measure defined on $\mathcal{N} \cup \{\varnothing\}$, the measure μ^τ constructed from τ by Method II satisfies

$$\mu^\tau(E) = \sup_{n \geqslant 0} \nu_{\mathcal{N}(n)}^\tau(E),$$

where $\nu_{\mathcal{N}(n)}^\tau$ is the measure constructed by Method I, from the restriction of τ to $\mathcal{N}(n) \cup \{\varnothing\}$, where

$$\mathcal{N}(n) = \mathcal{N} \backslash \{N_1, N_2. \ldots, N_n\}.$$

Proof. As $\mathcal{N}$ is enumerable, we may choose a sequence $L_1, L_2, \ldots$ of sets of $\mathcal{N}$ of positive diameter so that each set of $\mathcal{N}$ appears infinitely often. We then form a sequence $M_1, M_2, \ldots$ inductively by taking M_1 to be the maximal set of $\mathcal{N}$ containing L_1, and more generally taking M_{n+1}, for $n \geq 1$, to be M_n, if L_{n+1} occurs in the sequence $M_1, M_2, \ldots M_n$, and to be the maximal set of

$$\mathcal{N} \backslash \{M_1, M_2, \ldots, M_n\}$$

containing L_{n+1}, otherwise. Then, if a set N of $\mathcal{N}$ has positive diameter and is a proper subset of just m of the sets of $\mathcal{N}$ when we come to the $(m+1)$-st occurrence, say $L_l = N$, of N in the sequence $L_1, L_2, \ldots$, we must find N among the sets

$$M_1, M_2, \ldots, M_l.$$

So the sequence $M_1, M_2, \ldots$ exhausts all the sets of $\mathcal{N}$ of positive diameter. Further, if $1 \leq r < s$, it is clear that either M_s is properly contained in $M_r = M_{r+1} = \ldots = M_s$ or $M_r \cap M_s = \varnothing$. Taking $N_1, N_2, \ldots$ to be the sequence obtained from the sequence $M_1, M_2, \ldots$ by elimination of repetitions we obtain the first result.

Now suppose that for some positive λ we have

$$\mu^\tau(E) > \lambda.$$

By (12), we can choose $\delta > 0$ so that

$$\nu_\delta^\tau(E) > \lambda.$$

Let $D_1, D_2, \ldots$ be an enumeration of the maximal sets of the net $\mathcal{N}_\delta$ of sets of $\mathcal{N}$ with diameter not exceeding δ. Then

$$\Omega = \bigcup_{r=1}^{\infty} D_r.$$

So, by theorem 52, we can choose R so large that

$$\nu_\delta^\tau\left(E \cap \bigcup_{r=1}^{R} D_r\right) > \lambda.$$

There are at most a finite number of elements of $\mathcal{N}$ that properly contain one of the sets $D_1, D_2, \ldots, D_R$, and all these sets, necessarily having positive diameter, will occur in the sequence $N_1, N_2, \ldots, N_S$, provided we choose a sufficiently large integer S. Now, any set of $\mathcal{N}$ that meets one of the sets $D_1, D_2, \ldots, D_R$ is either contained in one of these sets and so has diameter not exceeding δ, or properly contains one of these sets and so is not in

$$\mathcal{N}(S) = \mathcal{N} \backslash \{N_1, N_2, \ldots, N_S\}.$$

As only sets of $\mathcal{N}(S)$ meeting $\bigcup_{r=1}^{r=R} D_r$ are concerned in the evaluation of

$$\nu^\tau_{\mathcal{N}(S)}\left(E \cap \bigcup_{r=1}^{R} D_r\right)$$

we conclude that

$$\nu^\tau_{\mathcal{N}(S)}\left(E \cap \bigcup_{r=1}^{R} D_r\right) \geqslant \nu^\tau_\delta\left(E \cap \bigcup_{r=1}^{R} D_r\right) > \lambda.$$

Hence

$$\sup_{n \geqslant 0} \nu^\tau_{\mathcal{N}(n)}(E) > \lambda.$$

But, if, for some positive λ,

$$\sup_{n \geqslant 0} \nu^\tau_{\mathcal{N}(n)}(E) > \lambda,$$

we can apply the same argument in reverse. We can choose S so large that

$$\nu^\tau_{\mathcal{N}(S)}(E) > \lambda.$$

Let $D_1, D_2, \ldots$ be an enumeration of the maximal sets of the net $\mathcal{N}(S)$. By theorem 52, we can choose R so large that

$$\nu^\tau_{\mathcal{N}(S)}\left(E \cap \bigcup_{r=1}^{R} D_r\right) > \lambda.$$

The sets of $\mathcal{N}$ that properly contain one of the sets $D_1, D_2, \ldots, D_R$ are finite in number and each has positive diameter. Choose $\delta > 0$ less than the smallest of these diameters. Now any cover of

$$E \cap \bigcup_{r=1}^{R} D_r$$

by sets of $\mathcal{N}_\delta$ can be reduced, by suitable omissions, to a cover of this set by sets of $\mathcal{N}_\delta$ that meet $\bigcup_{r=1}^{r=R} D_r$. But the sets of $\mathcal{N}$ that properly contain one of the sets $D_1, D_2, \ldots, D_r$ have diameter exceeding δ. So the reduced cover is by sets contained in one of $D_1, D_2, \ldots, D_R$ and so by sets of $\mathcal{N}(S)$. Hence

$$\nu^\tau_\delta\left(E \cap \bigcup_{r=1}^{R} D_r\right) \geqslant \nu^\tau_{\mathcal{N}(S)}\left(E \cap \bigcup_{r=1}^{R} D_r\right) > \lambda,$$

and

$$\mu^\tau(E) \geqslant \mu^\tau\left(E \cap \bigcup_{r=1}^{R} D_r\right) > \lambda.$$

We conclude that

$$\mu^\tau(E) = \sup_{n \geqslant 0} \nu^\tau_{\mathcal{N}(n)}(E).$$

7.7. We can now show that any compact set of positive measure with respect to a net measure, satisfying appropriate conditions, always has a compact subset of finite positive measure.

Theorem 54. *Let $\mathcal{N}$ be a net defined in a metric space Ω. Let μ^τ be a net measure constructed by Method II from a pre-measure τ defined on $\mathcal{N} \cup \{\varnothing\}$ and let ν_δ^τ, $\delta > 0$, be the corresponding auxiliary measures. Suppose that:*

(*a*) *τ takes only finite value on $\mathcal{N}$;*

(*b*) *no single point in Ω has infinite μ^τ-measure;*

(*c*) *If $\delta > 0$, and $K_1, K_2, \ldots$ is a decreasing sequence of compact sets with $\mu^\tau(\bigcap_{r=1}^{r=\infty} K_r) = 0$, then $\nu_\delta^\tau(K_r) \to 0$ as $r \to \infty$.*

Then, if K is a compact set of positive μ^τ-measure, K contains a compact subset of finite positive μ^τ-measure.

Proof. If k were a point of K with

$$\mu^\tau(\{k\}) > 0,$$

the condition (*b*) would ensure that

$$0 < \mu^\tau(\{k\}) < +\infty,$$

so that the set $\{k\}$ would be a compact subset of K of the required type. Thus we may suppose that, for each point k of K, we have

$$\mu^\tau(\{k\}) = 0. \tag{13}$$

Let $N_1, N_2, \ldots$ be an enumeration of the sets of $\mathcal{N}$ of positive diameter satisfying the condition of theorem 53, and adopt the notation of that theorem. By that theorem

$$\sup_{n \geqslant 0} \nu^\tau_{\mathcal{N}(n)}(K) = \mu^\tau(K) > 0.$$

So we can choose a non-negative integer m with

$$\nu^\tau_{\mathcal{N}(m)}(K) > 0. \tag{14}$$

As a preliminary step we construct a compact subset K^* of K that has positive μ^τ-measure and that meets each set of the form

$$N_n \quad \text{or} \quad \Omega \backslash N_n$$

with $n \geqslant 1$, in a closed set. Choose λ with

$$\nu^\tau_{\mathcal{N}(m)}(K) > \lambda > 0.$$

Take $K_0 = K$ and suppose that $K_0, K_1, \ldots, K_n$ are compact sets that have been chosen so that:

$$K_0 \supset K_1 \supset \ldots \supset K_n;$$

$$\nu^\tau_{\mathcal{N}(m)}(K_r) > \lambda \quad (r = 0, 1, \ldots, n);$$

$$K_r \cap N_r \quad \text{and} \quad K_r \cap (\Omega \backslash N_r) \quad \text{are compact} \quad (r = 1, 2, \ldots, n).$$

We proceed to an inductive choice of K_{n+1}. By condition (e) on our nets, the set N_{n+1} is an $\mathscr{F}_\sigma$-set, and further $\Omega\backslash N_{n+1}$ being the union of the countable system of those sets of $\mathscr{N}$ that do not meet N_{n+1} is also an $\mathscr{F}_\sigma$-set. Hence, by the increasing sets lemma (theorem 52), we can choose closed subsets, F_{n+1}, H_{n+1} say, of N_{n+1} and $\Omega\backslash N_{n+1}$ with

$$\nu^\tau_{\mathscr{N}(m)}(K_n \cap \{F_{n+1} \cup H_{n+1}\}) > \lambda.$$

Now taking
$$K_{n+1} = K_n \cap \{F_{n+1} \cup H_{n+1}\},$$

we satisfy our requirements with n replaced by $n+1$. We suppose the infinite sequence $K_0, K_1, \ldots$ chosen inductively in this way so that:

$$K_0 \supset K_1 \supset \ldots \supset K_r \supset \ldots;$$

$$\nu^\tau_{\mathscr{N}(m)}(K_r) > \lambda \quad (r = 0, 1, \ldots); \tag{15}$$

$$K_r \cap N_r \quad \text{and} \quad K_r \cap (\Omega\backslash N_r) \quad \text{are compact for} \quad r = 1, 2, \ldots.$$

Write
$$K^* = \bigcap_{r=0}^{\infty} K_r.$$

Then, for each $r \geqslant 1$, the sets

$$K^* \cap N_r = K^* \cap \{K_r \cap N_r\} = K^* \cap F_r,$$

$$K^* \cap \{\Omega\backslash N_r\} = K^* \cap [K_r \cap \{\Omega\backslash N_r\}] = K^* \cap H_r,$$

are compact. So our preliminary step will be complete, if we have

$$\mu^\tau(K^*) > 0. \tag{16}$$

Suppose we had
$$\mu^\tau(K^*) = 0.$$

Then, by our condition (c), we have

$$\nu^\tau_\delta(K_r) \to 0 \quad \text{as} \quad r \to \infty, \tag{17}$$

for each $\delta > 0$. But, provided we take δ with

$$0 < \delta < \min_{1 \leqslant r \leqslant m} d(N_r),$$

as we may, we have
$$\nu^\tau_{\mathscr{N}(m)}(E) \leqslant \nu^\tau_\delta(E), \tag{18}$$

for all sets E. But this together with (17) contradicts (15) for all sufficiently large r. Thus K^* satisfies our requirements.

By theorem 53 again, we can choose a new integer m with

$$\nu^\tau_{\mathscr{N}(m)}(K^*) > 0.$$

No
$$\nu^\tau_{\mathscr{N}(m)}\left(K^* \cap \bigcup_{r=m+1}^{n} N_r\right) \leqslant \sum_{r=m+1}^{n} \tau(N_r) < +\infty,$$

for each n. So, by theorem 52, we can choose an integer n with

$$0 < \nu^{\tau}_{\mathcal{N}(m)}\left(K^* \cap \bigcup_{r=m+1}^{n} N_r\right) < +\infty.$$

Replacing K^* by $K^* \cap \bigcup_{r=m+1}^{n} N_r$ we obtain a compact set K^* satisfying our previous requirements and having

$$0 < \nu^{\tau}_{\mathcal{N}(m)}(K^*) < +\infty.$$

Write
$$\nu^{\tau}_{\mathcal{N}(m)}(K^*) = \lambda.$$

Our next aim is to choose a decreasing sequence

$$K_m = K^*, K_{m+1}, \ldots, K_s, \ldots$$

of compact sets with

$$\nu^{\tau}_{\mathcal{N}(r)}(K_s) = \lambda \quad \text{for} \quad m \leqslant r \leqslant s.$$

Suppose that, for some $n \geqslant m$, such sets $K_m = K^*$, $K_{m+1}, \ldots, K_n$ have been chosen. Then

$$\nu^{\tau}_{\mathcal{N}(n+1)}(K_n) \leqslant \nu^{\tau}_{\mathcal{N}(n)}(K_n) = \lambda.$$

If
$$\nu^{\tau}_{\mathcal{N}(n+1)}(K_n) = \nu^{\tau}_{\mathcal{N}(n)}(K_n),$$

we take
$$K_{n+1} = K_n,$$

and have
$$\nu^{\tau}_{\mathcal{N}(r)}(K_s) = \lambda \quad \text{for} \quad m \leqslant r \leqslant s \leqslant n+1.$$

Otherwise
$$\nu^{\tau}_{\mathcal{N}(n+1)}(K_n) > \nu^{\tau}_{\mathcal{N}(n)}(K_n) = \lambda. \tag{19}$$

In this case it will be more difficult to define K_{n+1}. Since any set N of $\mathcal{N}(n)$ is either contained in N_{n+1} or in $\Omega \backslash N_{n+1}$, it follows immediately that

$$\nu^{\tau}_{\mathcal{N}(n)}(K_n) = \nu^{\tau}_{\mathcal{N}(n)}(K_n \cap N_{n+1}) + \nu^{\tau}_{\mathcal{N}(n)}(K_n \backslash N_{n+1}),$$

and

$$\nu^{\tau}_{\mathcal{N}(n+1)}(K_n) = \nu^{\tau}_{\mathcal{N}(n+1)}(K_n \cap N_{n+1}) + \nu^{\tau}_{\mathcal{N}(n+1)}(K_n \backslash N_{n+1}).$$

Further, as the set N_{n+1}, excluded from $\mathcal{N}(n)$ in forming $\mathcal{N}(n+1)$, covers no part of $K_n \backslash N_{n+1}$, we have

$$\nu^{\tau}_{\mathcal{N}(n+1)}(K_n \backslash N_{n+1}) = \nu^{\tau}_{\mathcal{N}(n)}(K_n \backslash N_{n+1}).$$

Hence
$$\nu^{\tau}_{\mathcal{N}(n+1)}(K_n \cap N_{n+1}) > \nu^{\tau}_{\mathcal{N}(n)}(K_n \cap N_{n+1}).$$

As these two measures of a subset of N_{n+1} differ, we must have

$$\nu^{\tau}_{\mathcal{N}(n+1)}(K_n \cap N_{n+1}) > \nu^{\tau}_{\mathcal{N}(n)}(K_n \cap N_{n+1}) = \tau(N_{n+1}). \tag{20}$$

Write

$$\eta = \nu^{\tau}_{\mathcal{N}(n)}(K_n) - \nu^{\tau}_{\mathcal{N}(n)}(K_n \backslash N_{n+1}) = \nu^{\tau}_{\mathcal{N}(n)}(K_n \cap N_{n+1}) = \tau(N_{n+1}).$$

Then, using (19)

$$\begin{aligned} \nu^{\tau}_{\mathcal{N}(n+1)}(K_n \cap N_{n+1}) &= \nu^{\tau}_{\mathcal{N}(n+1)}(K_n) - \nu^{\tau}_{\mathcal{N}(n)}(K_n \backslash N_{n+1}) \\ &> \nu^{\tau}_{\mathcal{N}(n)}(K_n) - \nu^{\tau}_{\mathcal{N}(n)}(K_n \backslash N_{n+1}) \\ &= \eta \geqslant 0. \end{aligned}$$

We concentrate our attention on the part of K_n in N_{n+1}. We make an inductive choice of a non-decreasing sequence of non-negative integers

$$r(0) = 0, \quad r(1), \quad r(2), \ldots,$$

of a disjoint sequence $\qquad L_1, L_2, \ldots$

of sets of $\mathcal{N}(n+1)$ of positive diameter, and a sequence

$$T_{r(0)} = N_{n+1}, \quad T_{r(1)}, \ldots$$

of sets of $\mathcal{N}(n)$, with the properties that

$$L_{r(s)+1}, L_{r(s)+2}, \ldots, L_{r(s+1)}, T_{r(s+1)},$$

are disjoint sets of $\mathcal{N}(n+1)$, properly contained in $T_{r(s)}$, $s = 0, 1, \ldots$; and that

$$\nu^{\tau}_{\mathcal{N}(n+1)}\left(K \cap \bigcup_{i=1}^{r(s)} L_i\right) \leqslant \eta < \nu^{\tau}_{\mathcal{N}(n+1)}\left(K_n \cap \left\{\left(\bigcup_{i=1}^{r(s)} L_i\right) \cup T_{r(s)}\right\}\right), \tag{21}$$

for $s = 0, 1, \ldots$. These conditions are trivially satisfied when $s = 0$, as $\bigcup_{i=1}^{i=r(0)} L_i = \varnothing$, so that (21) reduces to

$$\nu^{\tau}_{\mathcal{N}(n+1)}(\varnothing) \leqslant \eta < \nu^{\tau}_{\mathcal{N}(n+1)}(K_n \cap N_{n+1}).$$

Suppose that, for some $s \geqslant 0$, the integers $r(0), r(1), \ldots, r(s)$, the sets $L_1, L_2, \ldots, L_{r(s)}, T_{r(s)}$ have been chosen. Let $M_1, M_2, \ldots$ be an enumeration of the sets of $\mathcal{N}(n+1)$ that are properly contained in $T_{r(s)}$ and are maximal subject to this condition. Then $M_1, M_2, \ldots$ are disjoint sets with $T_{r(s)}$ as their union. By the increasing sets lemma (theorem 52),

$$\sup_R \nu^{\tau}_{\mathcal{N}(n+1)}\left(K_n \cap \left\{\left(\bigcup_{i=1}^{r(s)} L_i\right) \cup \left(\bigcup_{j=1}^{R} M_j\right)\right\}\right)$$

$$= \nu^{\tau}_{\mathcal{N}(n+1)}\left(K_n \cap \left\{\left(\bigcup_{i=1}^{r(s)} L_i\right) \cup T_{r(s)}\right\}\right) > \eta.$$

So we can choose R so that

$$\nu^{\tau}_{\mathcal{N}(n+1)}\left(K_n \cap \left\{\left(\bigcup_{i=1}^{r(s)} L_i\right) \cup \left(\bigcup_{j=1}^{R} M_j\right)\right\}\right)$$

$$\leqslant \eta < \nu^{\tau}_{\mathcal{N}(n+1)}\left(K_n \cap \left(\left\{\bigcup_{i=1}^{r(s)} L_i\right) \cup \left(\bigcup_{j=1}^{R+1} M_j\right)\right\}\right).$$

Let $L_{r(s)+1}, L_{r(s)+2}, \ldots, L_{r(s+1)}$ be an enumeration of those of the sets $M_1, M_2, \ldots, M_R$ that have positive diameter. Write $T_{r(s+1)} = M_{R+1}$. Any one-point sets among $M_1, M_2, \ldots, M_{R+1}$ have zero μ^τ-measure and so have zero $\nu^\tau_{\mathcal{N}(n+1)}$-measure and are $\nu^\tau_{\mathcal{N}(n+1)}$-measurable. Hence

$$\nu^\tau_{\mathcal{N}(n+1)}\left(K_n \cap \bigcup_{i=1}^{r(s+1)} L_i\right) \leqslant \eta < \nu^\tau_{\mathcal{N}(n+1)}\left(K_n \cap \left\{\left(\bigcup_{i=1}^{r(s+1)} L_i\right) \cup T_{r(s+1)}\right\}\right),$$

as required. In this way the inductive construction may be completed.

We now study the set

$$K_{n+1} = \{K_n \backslash N_{n+1}\} \cup \left\{\bigcap_{s=1}^{\infty} \left(K_n \cap \left\{\left(\bigcup_{i=1}^{r(s)} L_i\right) \cup T_{r(s)}\right\}\right)\right\}. \tag{22}$$

As K_n is a closed subset of K^* and as our preliminary construction ensures that

$$K^* \cap N, \quad K^* \backslash N$$

are closed for all N of $\mathcal{N}$, the sets

$$K_n \cap N = K_n \cap \{K^* \cap N\}, \qquad K_n \backslash N = K_n \cap \{K^* \backslash N\}$$

are closed for all N of $\mathcal{N}$. Hence, the set K_{n+1} defined by (22) is closed.

As

$$K_n \cap \left(\bigcup_{i=1}^{r(s)} L_i\right) \subset K_{n+1} \cap N_{n+1} \subset K_n \cap \left\{\left(\bigcup_{i=1}^{r(s)} L_i\right) \cup T_{r(s)}\right\},$$

for each $s \geqslant 0$, we have

$$\sup_s \nu^\tau_{\mathcal{N}(n+1)}\left(K_n \cap \left(\bigcup_{i=1}^{r(s)} L_i\right)\right) \leqslant \nu^\tau_{\mathcal{N}(n+1)}(K_{n+1} \cap N_{n+1})$$
$$\leqslant \inf_s \nu^\tau_{\mathcal{N}(n+1)}\left(K_n \cap \left\{\left(\bigcup_{i=1}^{r(s)} L_i\right) \cup T_{r(s)}\right\}\right).$$

By (21) we will have

$$\nu^\tau_{\mathcal{N}(n+1)}(K_{n+1} \cap N_{n+1}) = \eta,$$

as soon as we can prove that

$$\nu^\tau_{\mathcal{N}(n+1)}\left(K_n \cap \left\{\left(\bigcup_{i=1}^{r(s)} L_i\right) \cup T_{r(s)}\right\}\right)$$
$$- \nu^\tau_{\mathcal{N}(n+1)}\left(K_n \cap \left(\bigcup_{i=1}^{r(s)} L_i\right)\right) \to 0 \quad \text{as} \quad s \to \infty.$$

For this purpose it suffices to prove that

$$\nu^\tau_{\mathcal{N}(n+1)}(K_n \cap T_{r(s)}) \to 0 \quad \text{as} \quad s \to \infty. \tag{23}$$

Now, as K_n is closed and $K^* \cap T_{r(s)}$ is compact,

$$K_n \cap T_{r(s)} = K_n \cap (K^* \cap T_{r(s)})$$

is compact. By the strict inequality in (21), $K_n \cap T_{r(s)}$ is non-empty. So the decreasing sequence $K_n \cap T_{r(s)}$, $s = 0, 1, \ldots$ of non-empty compact sets has a non-empty intersection, T, say. Let t be any point of T. Now t belongs to sets of $\mathcal{N}$ of arbitrarily small diameter, while $T_{r(s+1)}$ is one of the maximal sets of $\mathcal{N}$ properly contained in $T_{r(s)}$ for $s = 0, 1, 2, \ldots$. As no set of $\mathcal{N}$ is contained in infinitely many sets of $\mathcal{N}$, it follows that

$$d(T_{r(s)}) \to 0 \quad \text{as} \quad s \to \infty.$$

Hence T has zero diameter and must be the singleton $\{t\}$. As $t \in K$ we have

$$\mu^\tau(\{t\}) = 0.$$

By condition (c), we conclude that, for each $\delta > 0$,

$$\nu_\delta^\tau(K_n \cap T_{r(s)}) \to 0 \quad \text{as} \quad s \to \infty. \tag{24}$$

As in the proof of (18), provided δ is sufficiently small,

$$\nu_{\mathcal{N}(n+1)}^\tau(E) \leqslant \nu_\delta^\tau(E),$$

for all sets E. So (23) follows from (24). Hence

$$\nu_{\mathcal{N}(n+1)}^\tau(K_{n+1} \cap N_{n+1}) = \eta,$$

and

$$\begin{aligned} \nu_{\mathcal{N}(n+1)}^\tau(K_{n+1}) &= \nu_{\mathcal{N}(n+1)}^\tau(K_{n+1} \cap N_{n+1}) + \nu_{\mathcal{N}(n+1)}^\tau(K_{n+1} \backslash N_{n+1}) \\ &= \eta + \nu_{\mathcal{N}(n+1)}^\tau(K_n \backslash N_{n+1}) \\ &= \eta + \nu_{\mathcal{N}(n)}^\tau(K_n \backslash N_{n+1}) \\ &= \nu_{\mathcal{N}(n)}^\tau(K_n) = \lambda. \end{aligned}$$

It remains to prove that

$$\nu_{\mathcal{N}(r)}^\tau(K_{n+1}) = \lambda \quad \text{for} \quad m \leqslant r \leqslant n. \tag{25}$$

Clearly $\qquad \nu_{\mathcal{N}(r)}^\tau(K_{n+1}) \leqslant \nu_{\mathcal{N}(r)}^\tau(K_n) \quad \text{for} \quad m \leqslant r \leqslant n.$

Suppose that for some r with $m \leqslant r \leqslant n$ we have

$$\nu_{\mathcal{N}(r)}^\tau(K_{n+1}) < \lambda,$$

then certainly

$$\nu_{\mathcal{N}(m)}^\tau(K_{n+1}) < \lambda. \tag{26}$$

So we can choose a sequence $L_1, L_2, \ldots$ of sets of $\mathcal{N}(m) \cup \{\varnothing\}$ with

$$K_{n+1} \subset \bigcup_{i=1}^\infty L_i, \qquad \sum_{i=1}^\infty \tau(L_i) < \lambda.$$

We may clearly suppose that all the sets $L_1, L_2, \ldots$ are disjoint. Let $L_1', L_2', \ldots$ be the finite or infinite sequence of those sets $L_1, L_2, \ldots$

that have a non-empty intersection with N_{n+1}, and let $L''_1, L''_2, \ldots$ be the remainder of the sequence. Suppose first that each set of the sequence $L'_1, L'_2, \ldots$ is properly contained in N_{n+1} and so is a set of $\mathcal{N}(n+1)$. As these sets $L'_1, L'_2, \ldots$ cover $K_{n+1} \cap N_{n+1}$ we must have

$$\sum_{i=1}^{\infty} \tau(L'_i) \geq \nu^{\tau}_{\mathcal{N}(n+1)}(K_{n+1} \cap N_{n+1}).$$

The strict inequality (26) excludes the possibility that $K_{n+1} = K_n$. So K_{n+1} must be obtained from K_n by the second method, and we must have

$$\nu^{\tau}_{\mathcal{N}(n+1)}(K_{n+1} \cap N_{n+1}) = \eta = \tau(N_{n+1}).$$

Thus

$$\tau(N_{n+1}) \leq \sum_{i=1}^{\infty} \tau(L'_i).$$

So the sets $N_{n+1}, L''_1, L''_2, \ldots$ cover K_{n+1} and

$$\tau(N_{n+1}) + \sum_{j=1}^{\infty} \tau(L''_j) \leq \sum_{i=1}^{\infty} \tau(L'_i) + \sum_{j=1}^{\infty} \tau(L''_j) < \lambda.$$

In the case when there is a set, say L'_1, of the sequence $L'_1, L'_2, \ldots$ that is not properly contained in N_{n+1}, the set L'_1 contains N_{n+1} and meets each set that meets N_{n+1}. So in this case the sequence $L'_1, L'_2, \ldots$ reduces to the single set L'_1. Thus, in each case, writing $L''_0 = N_{n+1}$ or $L''_0 = L'_1$, we have sets $L''_0, L''_1, \ldots$ of $\mathcal{N}(m)$, with

$$N_{n+1} \subset L''_0, \qquad K_{n+1} \subset \bigcup_{j=0}^{\infty} L''_j, \qquad \sum_{j=0}^{\infty} \tau(L''_j) < \lambda.$$

Since

$$\begin{aligned} K_n &= \{K_n \cap N_{n+1}\} \cup \{K_n \backslash N_{n+1}\} \\ &\subset N_{n+1} \cup \{K_{n+1} \backslash N_{n+1}\} \\ &\subset N_{n+1} \cup K_{n+1} \subset \bigcup_{j=0}^{\infty} L''_j, \end{aligned}$$

we have

$$\nu^{\tau}_{\mathcal{N}(m)}(K_n) < \lambda.$$

This contradiction to our choice of K_n, implies that (25) holds, and completes the proof that K_{n+1} satisfies our requirements.

We can now suppose that K_n is defined for all $n \geq m$. We write

$$K_0 = \bigcap_{n=m}^{\infty} K_m,$$

and show that K_0 is a compact subset of K with finite positive μ^{τ}-

measure. As $K_m, K_{m+1}, \ldots$, are compact subsets of K, it follows that K_0 is also a compact subset of K. For each integer n, if $l = \max(n, m)$ we have

$$\nu^\tau_{\mathcal{N}(n)}(K_0) \leqslant \nu^\tau_{\mathcal{N}(l)}(K_l) = \lambda.$$

So, by theorem 53, $\mu^\tau(K_0) = \sup_n \nu^\tau_{\mathcal{N}(n)}(K_0) \leqslant \lambda.$

Thus, K_0 has finite μ^τ-measure. As

$$\nu^\tau_{\mathcal{N}(m)}(K_r) = \lambda \quad (r = m, m+1, \ldots),$$

it follows, by precisely the argument used earlier to prove that K^* has positive μ^τ-measure, that K_0 has positive μ^τ-measure. This completes the proof of the theorem.

7.8. We now show how this result can be extended to Souslin-$\mathscr{F}$ sets in separable complete metric spaces when the net measure satisfies a slightly stronger condition. We recall (definition 30) that a decreasing sequence of sets $E_1, E_2, \ldots$ with intersection E converges strongly to E, if, for every open set G, with $E \subset G$, we have $E_r \subset G$ for all sufficiently large r.

***Theorem* 55.** *Let N be a net defined in a complete separable metric space Ω. Let μ^τ be a net measure constructed by Method II from a pre-measure τ defined on $\mathcal{N} \cup \{\varnothing\}$ and let ν^τ_δ, $\delta > 0$, be the corresponding auxiliary measures. Suppose that:*

(*a*) *τ takes only finite values on $\mathcal{N}$;*

(*b*) *no single point in Ω has infinite μ^τ-measure;*

(*c*) *if $\delta > 0$ and $\nu^\tau_\delta(E) = 0$ for any set E, then $\mu^\tau(E) = 0$;*

(*d*) *if $\delta > 0$, and $E_1, E_2, \ldots$ is a decreasing sequence of sets converging strongly to a set $E = \bigcap_{r=1}^{r=\infty} E_r$ with $\mu^\tau(E) = 0$, then $\nu^\tau_\delta(E_r) \to 0$ as $r \to \infty$.*

Then, if A is a Souslin-$\mathscr{F}$ set of positive μ^τ-measure, A contains a compact subset of finite positive μ^τ-measure. Further $\mu^\tau(A)$ is the supremum of the μ^τ-measures of these compact subsets of finite positive μ^τ-measure.

Proof. As $\mu^\tau(A) > 0$, we can choose $\delta > 0$ with

$$\nu^\tau_\delta(A) > 0.$$

By theorem 52, the strong form of the increasing sets lemma holds for ν^τ_δ-measure. Now, if $E_1, E_2, \ldots$ is any decreasing sequence of sets converging strongly to a set $E = \bigcap_{r=1}^{r=\infty} E_r$ with $\nu^\tau_\delta(E) = 0$, it follows from (*c*) that $\mu^\tau(E) = 0$, and then from (*d*) that $\nu^\tau_\delta(E) \to 0$ as $r \to \infty$. So the measure ν_δ^τ satisfies the conditions of corollary 3 to theorem 48, page 100, and, by that theorem, A must contain a compact set K of positive ν_δ^τ-measure. Then, necessarily, $\mu^\tau(K) > 0$.

Now, if $K_1, K_2, \ldots$ is any decreasing sequence of compact sets with $\mu^\tau(\bigcap_{r=1}^{r=\infty} K_r) = 0$, the decreasing sequence $K_1, K_2, \ldots$ necessarily converges strongly and, by condition (d),

$$\nu_\delta^\tau(K_r) \to 0 \quad \text{as} \quad r \to \infty,$$

for each $\delta > 0$. Hence the conditions of theorem 54 are all satisfied and K and so A contains a compact subset of finite positive μ^τ-measure.

Let λ be the supremum of the μ^τ-measures of the compact subsets K of A having finite positive μ^τ-measure. Suppose, if possible, that

$$\lambda < \mu^\tau(A).$$

Then we can choose an increasing sequence of compact subsets $K_1, K_2, \ldots$ of A with

$$0 < \mu^\tau(K_r) < \lambda,$$

and

$$\lim_{r\to\infty} \mu^\tau(K_r) = \lambda.$$

Now

$$A \backslash \left\{ \bigcup_{r=1}^{\infty} K_r \right\}$$

is a Souslin-$\mathscr{F}$ set with

$$\mu^\tau\left(A \backslash \left\{ \bigcup_{r=1}^{\infty} K_r \right\}\right) = \mu^\tau(A) - \lambda > 0,$$

as $\bigcup_{r=1}^{r=\infty} K_r$ is μ^τ-measurable. So, by the part of this theorem we have already proved, we can choose a compact subset K_0 of $A\backslash\{\bigcup_{r=1}^{r=\infty} K_r\}$ with finite positive μ^τ-measure. Hence

$$\mu^\tau\left(\bigcup_{r=0}^{\infty} K_r\right) = \mu^\tau(K_0) + \lambda,$$

and

$$\mu^\tau\left(\bigcup_{r=0}^{R} K_r\right) > \lambda,$$

for some sufficiently large R. Then $\bigcup_{r=0}^{r=R} K_r$ is a compact subset of finite μ^τ-measure too large for consistency with the definition of λ. This contradiction completes the proof of the theorem.

7.9. We now specialize the last theorem to yield a result for Hausdorff measures that can be approximated sufficiently closely by net measures.

***Theorem* 56.** *Let Ω be a complete separable metric space. Let h be a function of $\mathscr{H}_0$. Suppose that there exists a net $\mathscr{N}$ on Ω and a net measure μ^τ constructed by Method II from a pre-measure τ defined on $\mathscr{N} \cup \{\varnothing\}$ and corresponding auxiliary measures ν_δ^τ satisfying:*

(*a*) *τ takes only finite values on $\mathscr{N}$;*

(*b*) *there is a finite positive number M, such that for each $\delta > 0$,*

$$\nu_\delta^\tau(E) \leqslant M\mu_{\delta/M}^h(E), \tag{27}$$

and

$$\mu_\delta^h(E) \leqslant M\nu_{\delta/M}^\tau(E), \tag{28}$$

for all E of Ω.

Then if A is a Souslin-$\mathscr{F}$ set of positive μ^h-measure, A contains a compact subset of finite positive μ^h-measure. Further $\mu^h(A)$ is the supremum of the μ^h-measures of these compact subsets of finite positive μ^h-measure.

Proof. As $\mu^h(A) > 0$, it follows from (28) that $\mu^\tau(A) > 0$. We verify that the conditions (*a*)–(*d*) of theorem 55 are satisfied. Clearly (*a*) of theorem 55 is just (*a*) of this theorem. The condition (*b*) follows from (27). Again (*c*) follows from (28) together with the observation that when $\mu_{M\delta}^h(E) = 0$, then $\mu^h(E) = 0$ so that $\mu^\tau(E) = 0$ by (27). Now suppose that $\delta > 0$, and that $E_1, E_2, \ldots$ is a decreasing sequence of sets converging strongly to a set $E = \bigcap_{r=1}^{r=\infty} E_r$ with $\mu^\tau(E) = 0$. Then by (28) we have $\mu^h(E) = 0$. Suppose that $\epsilon > 0$ is given. Then we can cover E by a sequence $\{G_i\}$ of open sets with

$$E \subset \bigcup_{i=1}^{\infty} G_i, \qquad d(G_i) \leqslant \delta/M, \qquad \sum_{i=1}^{\infty} h(G_i) \leqslant \epsilon/M.$$

By the strong convergence of $E_1, E_2, \ldots,$ to E, we have

$$E_r \subset \bigcup_{i=1}^{\infty} G_i,$$

for all sufficiently large r. So, for such r,

$$\mu_{\delta/M}^h(E_r) \leqslant \epsilon/M.$$

Hence, by (27),

$$\nu_\delta^\tau(E_r) \leqslant \epsilon,$$

for all sufficiently large r. Thus

$$\nu_\delta^\tau(E_r) \to 0 \quad \text{as} \quad r \to \infty.$$

So condition (*d*) of theorem 55 is also satisfied. By that theorem, we conclude that A has a compact subset K of finite positive μ^τ-measure. By (27) and (28) it follows that K has finite positive μ^h-measure. By the argument of the last paragraph of the proof of theorem 55 we can easily complete the proof of this theorem.

Combining this theorem with theorem 49, with the result (3) of D. G. Larman (not proved here) and with theorem 50 one immediately obtains:

***Theorem* 57.** *Let Ω be a complete separable metric space. Let h be a continuous function of $\mathscr{H}_0$. Suppose that either*

(*a*) *Ω is n-dimensional Euclidean space, or*

(*b*) *Ω is compact and has zero (n)-measure for some positive integer n; or*

(*c*) *Ω is an ultra-metric space.*

If A is a Souslin-$\mathscr{F}$ set of positive μ^h-measure, A contains a compact subset of finite positive μ^h-measure. Further $\mu^h(A)$ is the supremum of the μ^h-measures of these compact subsets of A of finite positive μ^h-measure.

Proof. This is immediate.

Remarks. The key step in the proof of this theorem is the construction, pages 114–118, of subset K_{n+1} of K_n with

$$\nu^{\tau}_{\mathscr{N}(n+1)}(K_{n+1}) = \nu^{\tau}_{\mathscr{N}(n)}(K_{n+1}) = \nu^{\tau}_{\mathscr{N}(n)}(K_n);$$

the rest of the proof can be regarded as an elaborate manoeuvre designed to make this step possible and to exploit it to prove the theorem. As far as I know, no such step is possible without the use of a net measure of some sort.

D. G. Larman (1966*c*) shows that, if the function h corresponds to a dimension less than 1, and if the complete separable metric space Ω is connected, the same result can be obtained by a very different method, that depends on a selection process leading to a construction similar to that used in the construction of the Cantor ternary set.

The alternatives available in the statement of the theorem can be regarded as saying that the result is true when the space is not too large in cases (*a*) and (*b*) and when the metric structure of the space is not too complicated in case (*c*). Some such restrictions are essential. Roy O. Davies and C. A. Rogers (1969) have constructed a rather large compact metric space that has a rather complicated metric structure, that has infinite μ^h-measure for a continuous h of $\mathscr{H}_0$ but which contains no subsets of finite positive μ^h-measure.

§8 Sets of non-σ-finite measure

We shall first establish the simple result that, if a set E contains an uncountable family of disjoint μ-measurable subsets of positive μ-measure, for some measure μ, then E is of non-σ-finite μ-measure. The rest of the section will be concerned with the problem of the extent to which the converse result is true. We do not attempt to prove the strongest results known in this direction, but merely show that if h is a continuous function of $\mathscr{H}_0$, and if every compact subset of Ω with infinite μ^h-measure contains a compact subset of finite positive μ^h-measure, then every compact set of Ω of non-σ-finite μ^h-measure contains $\mathbf{c}$ disjoint compact subsets each of positive μ^h-measure.

8.1. We first prove

***Theorem* 58.** *If μ is a measure and a set E contains an uncountable family $\{E_\alpha\}_{\alpha \in A}$ of disjoint μ-measurable subsets E_α of positive μ-measure, then E is of non-σ-finite μ-measure.*

Proof. Suppose that the hypotheses are satisfied but that

$$E = \bigcup_{i=1}^{\infty} S_i,$$

where $\mu(S_i)$ is finite for $i = 1, 2, \ldots$. By the fundamental measure property

$$0 < \mu(E_\alpha) = \mu\left(\bigcup_{i=1}^{\infty} \{S_i \cap E_\alpha\}\right) \leqslant \sum_{i=1}^{\infty} \mu(S_i \cap E_\alpha),$$

for all α in A. So, for each α in A, there is a positive integer $i(\alpha)$ with

$$\mu(S_{i(\alpha)} \cap E_\alpha) > 0.$$

As a countable union of countable sets is countable, it follows that there is a positive integer i^* such that

$$\mu(S_{i^*} \cap E_\beta) > 0,$$

for all β in some uncountable subset B of A. Let $B(n)$ be the subset of B consisting of all β with

$$\mu(S_{i^*} \cap E_\beta) > 1/n.$$

As $B = \bigcup_{n=1}^{n=\infty} B(n)$, one of the sets $B(n)$, say $B(n^*)$, is uncountable. Let C be an infinite countable subset of $B(n^*)$. Then, by lemma 2 of Chapter 1,

$$\begin{aligned}\mu(S_{i^*}) &\geqslant \mu(S_{i^*} \cap \bigcup_{\gamma \in C} E_\gamma) \\ &= \sum_{\gamma \in C} \mu(S_{i^*} \cap E_\gamma) \\ &\geqslant \sum_{\gamma \in C} (1/n^*) = +\infty,\end{aligned}$$

contrary to our supposition that each set S_i has finite measure. This proves the theorem.

8.2. This result raises the question of whether or not a set of non-σ-finite measure necessarily contains such an uncountable family of disjoint measurable subsets of positive measure. One can only expect a positive answer to this question on the assumption that the original set is measurable. We can only establish a positive result in rather special circumstances. The following result is a weakened form of a result of Roy O. Davies (1968) that generalizes previous results of A. S. Besicovitch (1942) and (1956*b*), Roy O. Davies (1956*b*) and D. G. Larman (1967*d*). We discuss the relationship between this theorem and Davies' stronger result in the remarks after the statement of

***Theorem* 59.** *Let Ω be a metric space and let h be a continuous function of $\mathscr{H}_0$. Suppose that every compact subset of Ω with infinite μ^h-measure contains a subset of finite positive μ^h-measure. Then every compact set of Ω of non-σ-finite μ^h-measure contains a system of* **c** *disjoint compact subsets each of positive μ^h-measure.*

Remarks. By a refinement (indicated below) of the proof given below, the compact subsets can be chosen to have non-σ-finite μ^h-measure. When Ω is a compact metric space and h is a continuous function of $\mathscr{H}_0$, by results of Roy O. Davies (1956*b*) and M. Sion and D. Sjerve (1962) mentioned (but not proved) at the end of §6.2 of Chapter 2, p. 100, every Souslin-$\mathscr{F}$ set of Ω of non-σ-finite μ^h-measure contains a compact subset of non-σ-finite measure which will, provided the conditions of the theorem are satisfied, contain a system of **c** disjoint compact subsets of non-σ-finite μ^h-measure. This leads to the stronger form of Davies' result.

Proof. Let K be a compact set of Ω with non-σ-finite μ^h-measure. If $F_1, F_2, \ldots$ is any countable system of disjoint closed subsets of K, each finite positive μ^h-measure, then the set

$$K \setminus \left\{ \bigcup_{n=1}^{\infty} F_n \right\}$$

is a Souslin-$\mathscr{F}$ set of non-σ-finite μ^h-measure. By corollary 2 to theorem 48, p. 99, this set $K \setminus \{\cup F_n\}$ contains a closed set, F' say, of positive μ^h-measure. By our hypothesis, F' contains a subset, S, say, of finite positive μ^h-measure. Then there exists a $\mathscr{G}_\delta$-set, T say, containing S and with the same μ^h-measure. As $S \subset (T \cap F') \subset T$, it follows that $T \cap F'$ is a measurable subset of F' with finite positive measure. By theorem 27, p. 50, it follows that F' contains a closed subset, F^* say, of finite positive measure. Thus $K\backslash\{\cup F_n\}$ contains a closed subset F^* of finite positive measure.

By a transfinite inductive application of this argument we can choose a transfinite sequence

$$F_1, F_2, \ldots, F_\alpha, \ldots \quad (1 \leqslant \alpha < \omega_1), \tag{1}$$

where α runs through all countable ordinals, i.e. through all ordinals less than ω_1 the first uncountable ordinal, so that each set is a closed subset of K of finite positive measure and all sets of the system are disjoint.

It is convenient to work with the measure ν^h constructed by Method I from the pre-measure $h(G)$ defined on the class $\mathscr{G}$ of open sets. It follows immediately that a set E has $\mu^h(E) = 0$, if, and only if, $\nu^h(E) = 0$. Now as all the sets of the uncountable sequence

$$F_\alpha \quad (1 \leqslant \alpha < \omega_1),$$

have $\nu^h(F_\alpha) > 0$, there must be a $\delta > 0$ such that

$$\nu^h(F_\alpha) > \delta,$$

for uncountably many α with $1 \leqslant \alpha < \omega_1$. Choose such a $\delta > 0$ and let B be the set of those countable ordinals β with

$$\nu^h(F_\beta) > \delta.$$

Now consider the space $\mathscr{K}$ of non-empty closed subsets of K, with the Hausdorff metric introduced in §6.1, p. 90. By Blaschke's selection theorem (lemma §6.1, p. 91) $\mathscr{K}$ is a compact space under this metric, as K is compact.

Now any compact metric space has a countable base for its open sets. So by corollary 2 to theorem 34 (the Cantor–Bendixson theorem) the uncountable set $\{F_\beta\}_{\beta \in B}$ has an uncountable subset, $\{F_\gamma\}_{\gamma \in C}$ say, that is dense in itself.

We recall the idea of the distance separating two disjoint non-empty closed subsets A, B of a metric space, defined by

$$\inf_{\substack{a \in A \\ b \in B}} \rho(a, b).$$

We use $\eta(A, B)$ to denote this separating distance.

For any non-empty closed set A and any positive number r, we use $\mathrm{Bor}\,(A; r)$ to denote the 'bordered' set of all points b with $\rho(b, a) \leqslant r$ for some point a of A. Clearly $\mathrm{Bor}\,(A; r)$ is always a closed set, when A is closed.

If A, B are two disjoint compact sets, the corresponding bordered sets

$$\mathrm{Bor}(A; \tfrac{1}{3}\eta(A,B)), \qquad \mathrm{Bor}(B; \tfrac{1}{3}\eta(A,B))$$

are necessarily disjoint; for if c was a point common to these two sets, there would be points a, b in A and B with

$$\rho(c,a) \leqslant \tfrac{1}{3}\eta(A,B), \qquad \rho(c,b) \leqslant \tfrac{1}{3}\eta(A,B),$$

so that

$$\rho(a,b) \leqslant \tfrac{2}{3}\eta(A,B) < \eta(A,B),$$

contrary to the definition of $\eta(A,B)$.

Now suppose that F is any closed set chosen from the system $\{F_\gamma\}_{\gamma\in C}$ and that ρ_0 is any positive number. As $\{F_\gamma\}_{\gamma\in C}$ is dense in itself, there is a set, $F(1)$ say, of $\{F_\gamma\}_{\gamma\in C}$ with $\rho(F(1),F) < \frac{1}{2}\rho_0$. Then, clearly

$$\mathrm{Bor}(F(1);\rho) \subset \mathrm{Bor}(F;\rho_0),$$

for $0 < \rho \leqslant \frac{1}{2}\rho_0$. For symmetry of notation we write $F(0) = F$. Write

$$\rho_1 = \min\{\tfrac{1}{2}\rho_0, \tfrac{1}{3}\eta(F(0),F(1))\}.$$

Then, $F(i_1)$ belongs to $\{F_\gamma\}_{\gamma\in C}$,

$$\mathrm{Bor}(F(i_1);\, \rho_1) \subset \mathrm{Bor}(F;\, \rho_0),$$

$$\mathrm{Bor}(F(0);\, \rho_1) \cap \mathrm{Bor}(F(1);\rho_1) = \varnothing,$$

for $i_1 = 0$ or 1.

Applying this argument inductively, it is clear that for $k = 1, 2, \ldots$ we can choose a positive number $\rho_k \leqslant (\frac{1}{2})^k \rho_0$, and a family

$$F(i_1, i_2, \ldots, i_k) \quad (i_1, i_2, \ldots, i_k = 0 \quad \text{or} \quad 1),$$

of 2^k distinct points of $\{F_\gamma\}_{\gamma\in C}$, such that

$$\mathrm{Bor}(F(i_1, i_2, \ldots, i_k);\, \rho_k) \subset \mathrm{Bor}(F(i_1, i_2, \ldots, i_{k-1});\, \rho_{k-1}),$$

for

$$i_1, i_2, \ldots, i_k = 0 \quad \text{or} \quad 1,$$

and

$$\mathrm{Bor}(F(i_1, i_2, \ldots, i_k);\, \rho_k) \cap \mathrm{Bor}(F(j_1, j_2, \ldots, j_k);\, \rho_k) = \varnothing,$$

for $i_1, i_2, \ldots, i_k \neq j_1, j_2, \ldots, j_k$.

For each infinite sequence

$$\mathbf{i} = i_1, i_2, \ldots$$

of 0's and 1's write

$$F(\mathbf{i}) = \bigcap_{n=1}^{\infty} \mathrm{Bor}(F(\mathbf{i}|n);\, \rho_n).$$

As $F(\mathbf{i})$ is defined as the intersection of a decreasing sequence of non-empty closed subsets of the compact set K, it is clearly compact and non-empty. If $\mathbf{i}$ and $\mathbf{j}$ are different sequences of 0's and 1's, we can choose n with

$$\mathbf{i}|n \neq \mathbf{j}|n$$

with the consequence

$$F(\mathbf{i}) \cap F(\mathbf{j}) \subset \text{Bor}\,(F(\mathbf{i}|n);\, \rho_n) \cap \text{Bor}\,(F(\mathbf{j}|n);\, \rho_n) = \varnothing.$$

Thus we have a system of $\mathbf{c}$ disjoint compact non-empty subsets of K.

Now suppose that for some $\mathbf{i}$ we had

$$\nu^h(F(\mathbf{i})) < \delta.$$

Then we could choose a family $\{G_i\}$ of open sets with

$$F(\mathbf{i}) \subset \bigcup_{i=1}^{\infty} G_i, \qquad \sum_{i=1}^{\infty} h(G_i) < \delta.$$

As $\bigcup_{i=1}^{i=\infty} G_i$ is open, there would be an integer n so that

$$\text{Bor}\,(F(\mathbf{i}|n);\, \rho_n) \subset \bigcup_{i=1}^{\infty} G_i.$$

Then
$$F(\mathbf{i}|n) \subset \bigcup_{i=1}^{\infty} G_i, \qquad \sum_{i=1}^{\infty} h(G_i) < \delta,$$

so that $F(\mathbf{i}|n)$ is a set of $\{F_\gamma\}_{\gamma \in C}$ with

$$\nu^h(F(\mathbf{i}|n)) < \delta,$$

contrary to the choice of the sets $\{F_\beta\}_{\beta \in B}$. This completes the proof.

Remarks. The sets $F(\mathbf{i})$ produced in this way, in contrast to the sets $\{F_\alpha\}_{\alpha \in A}$, form a highly organized system: they are the points of a perfect set in the space $\mathscr{K}$; their union forms a compact subset of K. To obtain $\mathbf{c}$ disjoint compact subsets of non-σ-finite μ^h-measure it suffices to take the system of sets

$$\bigcup_{\mathbf{e}(\mathbf{i})=\mathbf{j}} F(\mathbf{i}),$$

$\mathbf{j}$ being a sequence of 0's and 1's and $\mathbf{e}(\mathbf{i})$ denoting the subsequence $i_2, i_4, \ldots$ of even co-ordinates of $\mathbf{i} = i_1, i_2, \ldots$.

3

APPLICATIONS OF HAUSDORFF MEASURES

§1 A survey of applications of Hausdorff measures

As I have only been able to devote a rather limited time to the preparation of this survey, it is very inadequate (like a half-hour visit to the British Museum), but I hope it will stimulate readers to explore some of the available literature.

Carathéodory's original development (1914) of his axiomatic approach to measure theory was clearly motivated by the desire to provide a framework for the linear and p-dimensional measures, with p integral, that he introduced in the same paper, as means of measurement of length and p-dimensional area. The relations between these length and area concepts and alternative concepts have been investigated by W. Gross (1918*a*, *b*), T. Estermann (1926), and later by S. Sherman (1942), H. Federer (1951, 1952), E. J. Mickle (1955) and others (see Federer 1969). The early investigations show that, if one is concerned with the assignment of an r-dimensional area to those sets of points that should have r-dimensional areas, then the Carathéodory method, equivalent to the use of the Hausdorff measure $\mu^{(r)}$ of Chapter 2, § 2, is as good as any other method. Although analysts tend to regard a surface as merely a set of points, geometers regard a surface as a more complicated object where some of the points have multiplicities exceeding unity. From this point of view, the surface area needs to be taken to be an integral of a multiplicity function defined on the surface (and zero off it) with respect to the appropriate Hausdorff measure. Much of the difficulty of the later investigations centres on the definition of the multiplicity function. The reader should consult H. Federer's recent book on geometric measure theory.

Hausdorff measures have played a dominant role in the development of surface area theory and, in particular, in the theory of minimal surfaces. We can only refer the reader to the series of papers by A. S. Besicovitch (1948*b*, *c*, 1949*a*–*c*, 1950, 1954*a*) and E. R. Reifenberg (1951, 1952*a*–*c*, 1955, 1960*a*, *b*, 1962, 1964) and to Federer's book (1969) and the many references given there.

In 1928 A. S. Besicovitch initiated a study of the geometrical properties of sets of points in the plane; this study has been pursued vigorously ever since. Besicovitch has mainly confined his attention to the case of sets of finite linear measure in the plane, and we shall follow his example in describing some of the results; but others have, not without great difficulty, extended his results to sets of finite positive $\mu^{(r)}$-measure in n-dimensional space with r a positive integer less than n. The upper and lower circular densities of a plane set E at a point n, not necessarily in E, with respect to the measure $\mu^{(1)}$ are defined to be the upper and lower limits of the ratio

$$\frac{\mu^{(1)}(E \cap D(x,r))}{d(D(x,r))}$$

as $r \to 0+$, $D(x,r)$ denoting the circular disc with centre x and radius r. The circular density is the common value of the upper and lower circular densities, when these do have a common value. Then at $\mu^{(1)}$-almost all points not in E, the circular density is zero; but, in contrast to the Lebesgue theory, it is not in general true that at $\mu^{(1)}$-almost all points of E the circular density has the value 1. The points of a set where the circular density has the value 1 are called the regular points of the set. Besicovitch shows that a $\mu^{(1)}$-measurable plane set E can be decomposed into two $\mu^{(1)}$-measurable sets R, I with

$$E = R \cup I, \qquad R \cap I = \varnothing,$$

so that R is regular at $\mu^{(1)}$-almost all of its points while I is regular at $\mu^{(1)}$-almost none of its points. Such sets R and I are called regular sets and irregular sets. The subsequent theory is largely concerned with establishing the contrasting properties of these two classes of sets; the density, tangential and projection properties of the two classes of sets turn out to be very different. Perhaps the most striking result of the theory is Besicovitch's embedding of each regular set in the union of a countable sequence of rectifiable arcs. The regular sets share most of the properties of the sets of points of the form

$$(\cos\theta, \sin\theta),$$

where θ lies in a Lebesgue measurable set of positive measure on the interval $[0, 2\pi]$. On the other hand, the irregular sets share most of the properties of the set $C \times C$, where C is a Cantor set on the unit interval chosen to ensure that $C \times C$ has finite positive $\mu^{(1)}$-measure. For generalizations of the general theory see J. M. Marstrand (1961) and W. C. Nemitz (1961). Although it is by no means easy to distinguish between (*a*) density, (*b*) tangential and (*c*) projection properties, we list some of the contributions to the geometrical theory of sets of points under these headings.

(*a*) Density properties are investigated by A. S. Besicovitch (1928, 1938, 1939, 1956*a*, 1957*a*, 1964*a*, 1967); G. Walker (1929); A. S. Besicovitch and G. Walker (1931); G. W. Morgan (1935); D. R. Dickinson (1939); J. M. Marstrand (1954*b*, *c*, 1964); E. J. Mickle and T. Rado (1958); A. S. Besicovitch and I. J. Schoenberg (1961); and D. G. Larman (1968).

(*b*) Tangential properties are investigated by A. S. Besicovitch in some of the works already mentioned and in (1956*b*, 1957*b*, 1960), J. Gillis (1935), E. R. Reifenberg (1951, 1952*a*, 1962, 1964), and J. M. Marstrand (1954*b*, 1961, 1964).

(*c*) Projection properties are investigated by A. S. Besicovitch in some of the works already mentioned, J. Gillis (1934), H. G. Eggleston (1958), and R. Kaufman (1968, 1969).

An account of much of this work will appear in A. S. Besicovitch's forthcoming book *Geometry of sets of points*.

A considerable amount of work has been done on the measures of cartesian product sets, of cylinders and on 'generalized cylinders'. Here the term cartesian product is used in the usual sense, and cylinder is used for the cartesian product of a set with an interval. If A and B are two sets that are naturally measured by two Hausdorff measures μ^f and μ^g, with f, g in $\mathscr{H}_0$, the general theory of cartesian product measures (see W. W. Bledsoe and A. P. Morse, 1955, for an account that discusses non-σ-finite measures) ensures that

$$\mu^f \times \mu^g(A \times B) = \mu^f(A) \times \mu^g(B),$$

with appropriate conventions when one of the factors on the right is infinite. But the cartesian product measure will not be a Hausdorff measure; and it is not immediately obvious how a Hausdorff measure appropriate for the study of $A \times B$ is to be found. If A and B are of finite positive μ^f- and μ^g-measures respectively, and if, for all $\epsilon > 0$, $\delta > 0$, it is possible to find a d with $0 < d < \delta$, and finite covers $I_1, I_2, \ldots, I_m$, and $J_1, J_2, \ldots, J_n$ of A and B with

$$d(I_i) = d \quad (i = 1, \ldots, m), \qquad \sum_{i=1}^{m} f(I_i) < \mu^f(A) + \epsilon,$$

$$d(J_j) = d \quad (j = 1, \ldots, n), \qquad \sum_{i=1}^{n} g(J_j) < \mu^g(B) + \epsilon,$$

the sets $I_i \times J_j$, $i = 1, \ldots, m$, $j = 1, \ldots, n$, cover $A \times B$, each has diameter at most $(\sqrt{2})\,d$, so that, on writing

$$h(t) = f(t/\sqrt{2}) \times g(t/\sqrt{2}),$$

we have
$$\sum_{i=1}^{m}\sum_{j=1}^{n} h(I_i \times J_j) \leqslant \sum_{i=1}^{m}\sum_{j=1}^{n} f(d)\,g(d)$$
$$= \left[\sum_{i=1}^{m} f(I_i)\right]\left[\sum_{j=1}^{n} g(J_j)\right]$$
$$\leqslant \{\mu^f(A)+\epsilon\}\{\mu^g(B)+\epsilon\};$$

and on these assumptions we have

$$\mu^h(A \times B) \leqslant \mu^f(A) \times \mu^g(B).$$

This suggests that the appropriate Hausdorff measure to use for the study of $A \times B$ is μ^h or the closely related measure $\mu^{f \cdot g}$. But the assumptions we have made can be badly wrong, it may happen that the diameters that lead to efficient covers of A lead only to inefficient covers of B, and vice versa; in this case we may expect the cartesian product set $A \times B$ to be of non-σ-finite $\mu^{f\cdot g}$-measure. This is indeed the case; while J. M. Marstrand (1954*a*) proves, under fairly general conditions, that, if $\mu^f(A) > 0$ and $\mu^g(B) > 0$ then $\mu^{f\cdot g}(A \times B) > 0$; it is only under rather special circumstances—for example, when B is an interval and f, g are sufficiently smooth—that it is possible to show that, if $\mu^f(A)$ and $\mu^g(B)$ are finite, then so is $\mu^{f\cdot g}(A \times B)$. For discussions of these problems and related problems see A. S. Besicovitch and P. A. P. Moran (1945), P. A. P. Moran (1949), H. G. Eggleston (1950*a*, 1954), G. Freilich (1950, 1956), S. J. Taylor (1952), J. M. Marstrand (1954*a*), D. J. Ward (1964, 1967), D. G. Larman and D. J. Ward (1966); D. G. Larman (1967*c*, *e*), and H. Wegmann (1969*a*, *b*).

A well-known result of H. Steinhaus (1920) says that, if E is a Lebesgue measurable set of positive Lebesgue measure on the line, then the distance set of all numbers $|y-x|$ with x, y numbers of E contains some interval including the number 0 as its left-hand end-point. Hausdorff measures have been used to investigate the ramifications of this and of related results by A. S. Besicovitch (1948*a*), A. S. Besicovitch and D. S. Miller (1948), H. G. Eggleston (1949), and A. S. Besicovitch and S. J. Taylor (1952).

The dimensional properties of the intersection of a linear set with its translates and linear transforms have been studied by P. A. P. Moran (1954) and by Roy O. Davies, J. M. Marstrand and S. J. Taylor (1960). Further the intersections of subsequences chosen from a given sequence of sets of positive Lebesgue measure have been studied by

P. Erdős, H. Kestelman and C. A. Rogers (1963) and by P. Erdős, and S. J. Taylor (1963).

Given a set E in a space Ω, one can ask whether there is a measure that gives a finite positive measure to E or to one of its subsets; and one can ask the same question for a class of sets, a class of spaces, a class of measures and a class of subsets. Negative answers, of varying degrees of sophistication, have been given to such questions by E. Best (1939), G. Choquet (1947), H. G. Eggleston (1954), and Roy O. Davies and C. A. Rogers (1969). Positive results have been discussed in Chapter 2.

By appeal to the geometrical duality between lines and points in a plane, measures can readily be assigned to sets of lines in the plane. Sets of lines are studied from a geometric-measure-theoretic point of view by A. S. Besicovitch (1964*b*) and Roy O. Davies (1965).

If a compact connected set K in a metric space Ω with metric ρ contains two distinct points a, b and if $\{G_i\}$ is any open cover of K, then $\bigcup_{i=1}^{i=N} G_i$ covers K for some N, and it is easy to show that

$$\sum_{i=1}^{N} d(G_i) \geqslant \rho(b, a).$$

From this we deduce that $\mu^{(1)}(K) \geqslant \rho(b, a) > 0$. More generally there are inequalities between the Hausdorff–Besicovitch dimensions of sets and their topological dimensions, see W. Hurewicz and H. Wallman (1941).

Hausdorff measures are often used to estimate the sizes of special sets that arise more or less naturally in branches of mathematics other than measure theory. We give some examples.

Curves defined parametrically by functions satisfying Lipschitz conditions are studied by A. S. Besicovitch and H. D. Ursell (1937) and by S. A. Kline (1945). Geographers have drawn attention to the analogy between the concepts of Hausdorff dimension and the fact that the apparent length of a river or coast-line increases as it is estimated by the study of more and more detailed maps.

Sets on the real line defined by the properties of the decimal expansions (or generalized decimal expansions) of their members have been discussed by E. Borel (1909), F. Hausdorff (1919), A. S. Besicovitch (1929, 1934*a*), E. Best (1940*a*, *b*, 1941, 1942), H. G. Eggleston (1952), A. Gierl (1959), J. Cigler (1961), W. A. Beyer (1962), J. R. Kinney (1958), P. Billingsley (1960, 1961), and M. Smorodinski (1968). The last three of these authors are concerned with sets of real numbers that are of interest in information theory.

Sets on the real line defined by the properties of the rational approximations to their members or by the properties of their continued fraction expansions have been discussed by E. Borel (1895), V. Jarník (1928), A. S. Besicovitch (1934*c*), I. J. Good (1941), H. G. Eggleston (1952), C. A. Rogers (1964), J. R. Kinney and T. S. Pitcher (1965/6, 1966*a*), K. E. Hirst (1970). See § 2 below for an account of a small part of this work.

In the mathematical theory of Brownian motion in n-dimensional Euclidean space, and in the theory of similar stochastic processes, a probability measure is introduced into the space of maps of the real half-line $[0, +\infty)$ into R^n taking 0 to the origin O of R^n. The image of $[0, +\infty)$ in R is called the path of the map. Many questions can be asked about the likely behaviour of the paths. By use of the general probabilistic theorem, called the law of 0 or 1, it is often easy to give the answer: *either* with probability 1 the path will have the given property *or* with probability 1 the path will not have the given property; but in many such cases it is very much more difficult to decide which of these alternative (mutually contradictory) conclusions will hold. In the case of Brownian motion the path is almost surely continuous and also almost surely non-differentiable at each of its points. Further the image of each finite segment $[0, t]$ almost surely has non-σ-finite $\mu^{(1)}$-measure. Similar questions have been asked about more general Hausdorff measures of the paths, of the set of $[0, +\infty)$ that maps onto the origin O, and the set of double, triple and multiple points of the paths. Many such questions have been answered by P. Lévy (1948, 1951, 1954), A. Dvoretzky, P. Erdős and S. Kakutani (1950, 1954), A. Dvoretzky, P. Erdős, S. Kakutani and S. J. Taylor (1957), A. Dvoretzky and S. Kakutani (1958), S. J. Taylor (1953, 1955*b*, 1964, 1966, 1967), P. Erdős and S. J. Taylor (1960*a*, *b*, 1961), R. M. Blumenthal and R. K. Getoor (1960*a*, *b*, 1961, 1962, 1964), Z. Ciesielski and S. J. Taylor (1962), D. Ray (1963), and S. J. Taylor and J. G. Wendel (1966), and J. Hawkes (1970).

The concept of the generalized capacity of a set is in some ways analogous to the concept of the Hausdorff measure of a set. Indeed there are some definite relations between the two concepts. These relations and the degrees of interdependence and independence between the two concepts have been investigated by S. Kametani (1942, 1945), V. Erohin (1958), L. Carleson (1958), and S. J. Taylor (1961).

A well-known result of Vitali (proved by a Vitali argument) shows that it is possible to choose a countable sequence $\{D_i\}_{i=1}^{i=\infty}$ of disjoint open circular discs, contained in a given disc D_0, and such that the

Lebesgue measure of the residual set $D_0\backslash\{\bigcup_{i=1}^{i=\infty} D_i\}$ is zero. This leads to the question 'How small can the residual set be made?' H. G. Eggleston (1953a) gives a precise answer to an analogous question on the packing of equilateral triangles; he shows that if a countable sequence $\{T_i\}_{i=1}^{i=\infty}$ of disjoint open equilateral triangles, all with the same orientation, lie in an equilateral triangle of the opposite orientation, then the residual set must have positive μ^h-measure, but will in a special case have finite μ^h-measure, with

$$h(t) = t^{(\log 3)/(\log 2)}.$$

Estimates for the fractional dimensions of the residual set in the problems of packings of circular discs and spherical balls are made by K. E. Hirst (1967) and D. G. Larman (1966a, b, 1967a). See also Z. A. Melzak (1966).

A number of other rather special geometrical investigations have exploited the theory of Hausdorff measures. P. Erdős (1946) investigates the set of points x in Euclidean n-space with the property that the system of points of a given closed set that are at the least possible distance from x is of linear dimension at least k, with $1 \leqslant k \leqslant n$. S. J. Taylor (1955) investigates the sets on a line that have no three points forming an arithmetic progression. D. J. Ward (1970a, b) has shown that, for each n, there is a plane set of Hausdorff dimension less than 2 that contains all n-gons; his results have been refined by Roy O. Davies (197?). T. J. McMinn (1960) investigates the set of directions that can be taken by line segments lying on the surface of a 3-dimensional convex body. A. S. Besicovitch (1963a, b) gives a simpler proof of McMinn's result and also studies the set of singularities of a certain type that can occur on the surface of a 3-dimensional convex body. G. Ewald, D. G. Larman and C. A. Rogers (1970) extend McMinn's results to r-flats on an n-dimensional convex body with $1 \leqslant r \leqslant n-1$.

P. Erdős (1940) studies the set of rational points in Hilbert space.

A. F. Beardon (1965a, b, 1966) studies the set of singular points of certain properly discontinuous transformation groups in N-dimensional space.

Let us turn from the study of sets to the study of functions. Certain differentiation properties of functions are investigated by A. S. Besicovitch (1934b) and by J. Ravetz (1954). R. Salem (1951) constructs monotonic functions whose Fourier transforms have certain dimensional properties. Functions that are continuous but not absolutely continuous are studied by C. A. Rogers and S. J. Taylor (1959, 1962, 1963), see also §3 below. J. R. Kinney (1960) studies the properties of

Minkowski's monotonic function that maps the rationals onto the real quadratic irrationals. J. R. Kinney and T. S. Pitcher (1964) study the support of a plane distribution function defined by a random process of giving weights to the two halves produced by repeated bisections of the unit square.

§2 Sets of real numbers defined in terms of their expansions into continued fractions

This section does not attempt to give a detailed account of the work under this general heading mentioned in §1, it merely gives an account of one of the results obtained by Jarník (1928) in his pioneering investigation.

2.1. We must start with a concise account of the elementary results in the theory of continued fractions that we shall need. Any real number x can be expressed uniquely in the form

$$x_0 = a_0 + \theta_1,$$

with a_0 integral and θ_1 real with $0 \leqslant \theta_1 < 1$. If x_0 is not an integer we have $\theta_1 \neq 0$ and there is a unique x_1 with

$$x_0 = a_0 + (1/x_1),$$

and $x_1 > 1$. Now, in the same way, if x_1 is not an integer, x_1 has a unique representation in the form

$$x_1 = a_1 + (1/x_2),$$

with a_1 integral and x_2 real with $x_2 > 1$. Unless this process terminates with an integer $x_n \geqslant 2$, it can be continued indefinitely. By induction we have

$$x_0 = a_0 + (1/[a_1 + (1/[a_2 + \ldots (1/[a_{n-1} + (1/x_n)]) \ldots])]),$$

or in the notation usually employed in the theory of continued fractions

$$x_0 = a_0 + \frac{1}{a_1 +}\,\frac{1}{a_2 +} \cdots \frac{1}{a_{n-1} +}\,\frac{1}{x_n}, \tag{1}$$

for all positive integers n, or for all positive integers n up to the stage when the process terminates.

If the expansion terminates, it is clear that x_0 is rational; if x_0 is rational, it is easy to verify that each x_n is a rational with a strictly smaller denominator than x_{n-1}, until the process terminates, as it clearly must after a finite number of steps. The integers $a_0, a_1, a_2, \ldots$

are called the partial quotients of the expansion. The standard results that we shall need from the theory of continued fractions are summarized in the following theorem; for more detailed accounts see G. H. Hardy and E. M. Wright (1954) or H. Davenport (1952).

Theorem 60. *Let $a_0, a_1, a_2, \ldots$ be a sequence of integers with $a_r \geqslant 1$ for $r \geqslant 1$. Let $p_n, q_n, n = -1, 0, 1, \ldots$ be defined by the initial conditions*

$$p_{-1} = 1, \quad q_{-1} = 0, \quad p_0 = a_0, \quad q_0 = 1, \tag{2}$$

and the recurrence relations

$$p_n = a_n p_{n-1} + p_{n-2}, \quad q_n = a_n q_{n-1} + q_{n-2} \quad (n = 1, 2, \ldots). \tag{3}$$

Then the sequence p_n/q_n, $n = 1, 2, \ldots$ converges to an irrational real number x_0 that yields the expansion

$$x_0 = a_0 + \frac{1}{a_1+}\frac{1}{a_2+}\ldots, \tag{4}$$

when the expansion process described above is applied.

Further, for each integer $n \geqslant 1$, and all real numbers x, y with $x > 1$, $y > 1$, we have:

$$p_n q_{n-1} - p_{n-1} q_n = (-1)^{n-1}; \tag{5}$$

$$a_0 + \frac{1}{a_1+}\frac{1}{a_2+}\cdots\frac{1}{a_n+}\frac{1}{x} = \frac{xp_n + p_{n-1}}{xq_n + q_{n-1}}; \tag{6}$$

$$\left|\left\{a_0 + \frac{1}{a_1+}\frac{1}{a_2+}\cdots\frac{1}{a_n+}\frac{1}{y}\right\} - \left\{a_0 + \frac{1}{a_1+}\frac{1}{a_2+}\cdots\frac{1}{a_n+}\frac{1}{x}\right\}\right|$$
$$= \frac{|y-x|}{(xq_n + q_{n-1})(yq_n + q_{n-1})}; \tag{7}$$

the number

$$a_0 + \frac{1}{a_1+}\frac{1}{a_2+}\cdots\frac{1}{a_n+}\frac{1}{x} \tag{8}$$

lies in the interior of the interval with end points

$$\frac{p_n}{q_n}, \qquad \frac{p_n + p_{n-1}}{q_n + q_{n-1}}, \tag{9}$$

and length

$$\frac{1}{q_n(q_n + q_{n-1})}; \tag{10}$$

$$q_n \geqslant (\tfrac{1}{2} + \tfrac{1}{2}\sqrt{5})^{n-1}; \tag{11}$$

and

$$x_0 - \frac{p_n}{q_n} = \frac{1}{q_n(x_{n+1}q_n + q_{n-1})}, \tag{12}$$

$x_{n+1} > 1$ being obtained by application of the expansion process to x_0.

Proof. Suppose $x > 1$. By the initial conditions (2) we have the identity

$$a_0 + \frac{1}{x} = \frac{a_0 x + 1}{x} = \frac{p_0 x + p_{-1}}{q_0 x + q_{-1}}.$$

But, if we have established the identity

$$a_0 + \frac{1}{a_1 +}\,\frac{1}{a_2 +} \cdots \frac{1}{a_n}\,\frac{1}{x} = \frac{x p_n + p_{n-1}}{x q_n + q_{n-1}}$$

for some $n \geqslant 1$, we have

$$\begin{aligned} a_0 + \frac{1}{a_1 +}\,\frac{1}{a_2 +} \cdots \frac{1}{a_n +}\,\frac{1}{a_{n+1} +}\,\frac{1}{x} \\ &= a_0 + \frac{1}{a_1 +}\,\frac{1}{a_2 +} \cdots \frac{1}{a_n +}\,\frac{1}{[a_{n+1} + x^{-1}]} \\ &= \frac{[a_{n+1} + x^{-1}]\,p_n + p_{n-1}}{[a_{n+1} + x^{-1}]\,q_n + q_{n-1}} \\ &= \frac{[a_{n+1} p_n + p_{n+1}]\,x + p_n}{[a_{n+1} q_n + q_{n-1}]\,x + q_n} \\ &= \frac{p_{n+1} x + p_n}{q_{n+1} x + q_n}. \end{aligned}$$

Thus the result (6) follows by induction.

Again, from the initial conditions (2), we have

$$p_0 q_{-1} - p_{-1} q_0 = a_0 . 0 - 1 . 1 = (-1)^{-1},$$

and from the recurrence relations (3),

$$\begin{aligned} p_{n+1} q_n - p_n q_{n+1} &= [a_{n+1} p_n + p_{n-1}]\,q_n - p_n [a_{n+1} q_n + q_{n-1}] \\ &= -(p_n q_{n-1} - p_{n-1} q_n), \end{aligned}$$

for $n \geqslant 0$. So the result (5) follows by induction.

From the initial conditions $q_{-1} = 0$, $q_0 = 1$ and the condition $a_1 \geqslant 1$, we have

$$q_0 = 1 > (\tfrac{1}{2} + \tfrac{1}{2}\sqrt{5})^{-1}, \qquad q_1 = a_1 \geqslant 1 = (\tfrac{1}{2} + \tfrac{1}{2}\sqrt{5})^{0}.$$

But once we know that

$$q_n \geqslant (\tfrac{1}{2} + \tfrac{1}{2}\sqrt{5})^{n-1}, \qquad q_{n-1} \geqslant (\tfrac{1}{2} + \tfrac{1}{2}\sqrt{5})^{n-2},$$

for some $n \geqslant 1$, we deduce that

$$\begin{aligned} q_{n+1} &= a_{n+1} q_n + q_{n-1} \\ &\geqslant (\tfrac{1}{2}+\tfrac{1}{2}\sqrt{5})^{n-1} + (\tfrac{1}{2}+\tfrac{1}{2}\sqrt{5})^{n-2} \\ &= (1\tfrac{1}{2}+\tfrac{1}{2}\sqrt{5})(\tfrac{1}{2}+\tfrac{1}{2}\sqrt{5})^{n-2} \\ &= (\tfrac{1}{2}+\tfrac{1}{2}\sqrt{5})^{n}. \end{aligned}$$

The result (11) follows by induction.

Using (5), we have

$$\frac{p_n}{q_n} - \frac{p_{n-1}}{q_{n-1}} = \frac{p_n q_{n-1} - p_{n-1} q_n}{q_n q_{n-1}} = \frac{(-1)^{n-1}}{q_n q_{n-1}},$$

for $n \geqslant 0$. By (11), we have

$$\left| \frac{p_n}{q_n} - \frac{p_{n-1}}{q_{n-1}} \right| \leqslant (\tfrac{1}{2}+\tfrac{1}{2}\sqrt{5})^{3-2n},$$

for $n \geqslant 2$. So the sequence $\{p_n/q_n\}$ is a Cauchy sequence converging to some real number x_0 say. Further, as the terms of the series

$$\sum_{r=0}^{\infty} \left\{ \frac{p_r}{q_r} - \frac{p_{r-1}}{q_{r-1}} \right\} = \sum_{r=0}^{\infty} \frac{(-1)^{r-1}}{q_r q_{r-1}}$$

alternate in sign and decrease numerically, the sum x_0 of the series lies strictly between any pair of consecutive terms. Thus, for each $n \geqslant 1$, x_0 lies strictly between

$$\frac{p_{n-1}}{q_{n-1}} \quad \text{and} \quad \frac{p_n}{q_n}.$$

Now let $b_0, b_1, \ldots$ be the sequence, finite or infinite, of integers, that is obtained when x_0 is expanded by the continued fraction process. As x_0 lies strictly between

$$\frac{p_0}{q_0} = a_0 \quad \text{and} \quad \frac{p_1}{q_1} = a_0 + \frac{1}{a_1},$$

we must have $b_0 = a_0$, and x_0 is not an integer. Suppose that, for some integer $n \geqslant 0$, we have shown that

$$b_0 = a_0, \quad b_1 = a_1, \ldots, b_n = a_n,$$

and that x_n is not an integer.

Then

$$x_0 = a_0 + \frac{1}{a_1+}\,\frac{1}{a_2+}\cdots\frac{1}{a_n+}\,\frac{1}{x_{n+1}},$$

and x_0 lies strictly between

$$\frac{p_{n+1}}{q_{n+1}} \quad \text{and} \quad \frac{p_{n+2}}{q_{n+2}}.$$

Now, for $x > 0$, the function

$$a_0 + \frac{1}{a_1+}\,\frac{1}{a_2+}\cdots\frac{1}{a_n+}\,\frac{1}{x} = \frac{p_n x + p_{n-1}}{q_n x + q_{n-1}}$$

is a monotonic function of x taking the value p_{n+1}/q_{n+1} when $x = a_{n+1}$ and the value p_{n+2}/q_{n+2} when $x = a_{n+1} + (1/a_{n+2})$. So the function takes the value x_0 for just one positive value of x, and this value lies between a_{n+1} and $a_{n+1} + (1/a_{n+2})$. Hence we must have

$$a_{n+1} < x_{n+1} < a_{n+1} + \frac{1}{a_{n+2}}.$$

Consequently $b_{n+1} = a_{n+1}$ and x_{n+1} is not an integer. It follows by induction that x_0 has the continued fraction expansion

$$a_0 + \frac{1}{a_1+}\,\frac{1}{a_2+}\cdots,$$

and that x_0 is irrational as this expansion does not terminate.

The outstanding parts of the theorem can now be readily obtained. By (6), if $x > 1, y > 1$, we have

$$\begin{aligned}\left\{a_0 + \frac{1}{a_1+}\,\frac{1}{a_2+}\cdots\frac{1}{a_n+}\,\frac{1}{y}\right\} - \left\{a_0 + \frac{1}{a_1+}\,\frac{1}{a_2+}\cdots\frac{1}{a_n+}\,\frac{1}{x}\right\} \\ = \frac{yp_n + p_{n-1}}{yq_n + q_{n-1}} - \frac{xp_n + p_{n-1}}{xq_n + q_{n-1}} \\ = \frac{(-1)^{n-1}(y-x)}{(xq_n + q_{n-1})(yq_n + q_{n-1})}, \qquad (13)\end{aligned}$$

on using (5), so that (7) follows. The result that the number (8) lies in the interior of the interval (9) follows immediately from (6). That the length of the interval (9) is (10) follows immediately from (5). The result (12) follows immediately from the identity (13) on taking $y = x_{n+1}$ and letting x tend to $+\infty$.

2.2. The theory of Diophantine Approximation tells us that the 'best' rational approximations to an irrational number are given by the rational numbers p_n/q_n $(n = 0, 1, 2, \ldots)$, provided by its continued fraction expansion. The rational number p_n/q_n is a best rational approximation in the sense that there is no rational number p/q with

$$\left|x_0 - \frac{p}{q}\right| < \left|x_0 - \frac{p_n}{q_n}\right| \quad (0 < q < q_n).$$

Further, as can be seen from (12), the rational number p_n/q_n will be an especially good rational approximation, if, and only if, x_{n+1}, or equivalently a_{n+1}, is especially large. So the irrationals x_0 with no especially good rational approximations are those where the partial quotient a_n takes only the values 1 and 2 for $n \geqslant 1$. Our aim will be to study the set of these irrational numbers. There is no loss in taking $a_0 = 0$, so we shall do this.

Throughout the rest of this section P will be used to denote the set of all irrational x_0 whose continued fraction expansion

$$x_0 = a_0 + \frac{1}{a_1 +}\,\frac{1}{a_2 +}\cdots,$$

satisfies the condition

$$a_0 = 0, \quad a_i = 1 \quad \text{or} \quad 2 \quad (\text{for} \quad i = 1, 2, \ldots).$$

Our aim will be to estimate the size of this set P. We do this by studying intervals suggested by (8) and (9) of theorem 60. For any set of integers $a_1, a_2, \ldots, a_n$ with $a_i = 1$ or 2, $1 \leqslant i \leqslant n$, let $I(a_1, a_2, \ldots, a_n)$ be the open interval with end points

$$\frac{p_n}{q_n} = \frac{1}{a_1 +}\,\frac{1}{a_2 +}\cdots\frac{1}{a_n}, \qquad \frac{p_n + p_{n-1}}{q_n + q_{n-1}} = \frac{1}{a_1 +}\,\frac{1}{a_2 +}\cdots\frac{1}{a_n +}\,\frac{1}{1},$$

$p_n, q_n, p_{n-1}, q_{n-1}$ being defined as in theorem 60, but with a_0 put equal to zero. Let $l(a_1, a_2, \ldots, a_n)$ denote the length of $I_0(a_1, a_2, \ldots, a_n)$. Then, by (10) of theorem 60,

$$l(a_1, a_2, \ldots, a_n) = \frac{1}{q_n(q_n + q_{n-1})}.$$

Note that, by (8) and (9) of theorem 60, the set P is covered by the system of 2^n intervals

$$I(a_1, a_2, \ldots, a_n) \quad (a_1, a_2, \ldots, a_n = 1 \quad \text{or} \quad 2),$$

for each fixed $n \geqslant 0$. The key to our estimate of the size of P will be the following lemmas of Jarník.

Lemma 1. *If $n \geqslant 0$ we have*

$$\{l(a_1, a_2, \ldots, a_n, 1)\}^s + \{l(a_1, a_2, \ldots, a_n, 2)\}^s \geqslant \{l(a_1, a_2, \ldots, a_n)\}^s, \qquad (14)$$

if $s = \frac{1}{3}$, and the reverse inequality holds if $s = \frac{2}{3}$.

Proof. We study the ratio

$$R = [\{l(a_1, a_2, \ldots, a_n, 1)\}^s + \{l(a_1, a_2, \ldots, a_n, 2)\}^s]/[\{l(a_1, a_2, \ldots, a_n)\}^s].$$

We have

$$l(a_1, a_2, \ldots, a_n, a_{n+1}) = \frac{1}{q_{n+1}(q_{n+1}+q_n)}$$

$$= \frac{1}{(a_{n+1}q_n + q_{n-1})(\{a_{n+1}+1\}q_n + q_{n-1})}$$

$$= \frac{1}{q_n^2(\beta + a_{n+1} - 1)(\beta + a_{n+1})}$$

on writing $\beta = (q_n + q_{n-1})/q_n.$

Then $1 < \beta < 2$ and

$$l(a_1, a_2, \ldots, a_n) = \frac{1}{q_n(q_n + q_{n-1})} = \frac{1}{q_n^2 \beta}.$$

Hence $$R = \left\{\frac{1}{\beta+1}\right\}^s + \left\{\frac{\beta}{(\beta+1)(\beta+2)}\right\}^s.$$

When $s = \frac{1}{3}$ we have

$$R = \left\{\frac{1}{\beta+1}\right\}^{\frac{1}{3}} + \left\{\frac{\beta}{2+3\beta+\beta^2}\right\}^{\frac{1}{3}}$$

$$> \left(\frac{1}{2+1}\right)^{\frac{1}{3}} + \left(\frac{1}{2+3\,.\,2+2^2}\right)^{\frac{1}{3}}$$

$$= (\tfrac{1}{3})^{\frac{1}{3}} + (\tfrac{1}{12})^{\frac{1}{3}} > (\tfrac{8}{27})^{\frac{1}{3}} + (\tfrac{1}{27})^{\frac{1}{3}} = 1.$$

When $s = \frac{2}{3}$ we have

$$R = \left\{\frac{1}{\beta+1}\right\}^{\frac{2}{3}} + \left\{\frac{\beta}{5\beta+1+(\beta-1)^2}\right\}^{\frac{2}{3}}$$

$$< (\tfrac{1}{2})^{\frac{2}{3}} + (\tfrac{1}{5})^{\frac{2}{3}}$$

$$< (\tfrac{64}{125})^{\frac{2}{3}} + (\tfrac{27}{125})^{\frac{2}{3}}$$

$$= (\tfrac{4}{5})^2 + (\tfrac{3}{5})^2 = 1.$$

This proves the lemma.

Lemma 2. *If irrationals x, y in P have continued fraction expansions*

$$x = \frac{1}{a_1+}\,\frac{1}{a_2+}\cdots, \qquad y = \frac{1}{b_1+}\,\frac{1}{b_2+}\cdots,$$

with $a_1 = b_1, a_2 = b_2, \ldots, a_n = b_n$, but $a_{n+1} \neq b_{n+1}$, then

$$\tfrac{1}{27} l(a_1, a_2, \ldots, a_n) < |x-y| < l(a_1, a_2, \ldots, a_n). \tag{15}$$

Proof. As $a_{n+1} = 1$ or 2 and $b_{n+1} = 1$ or 2 and $a_{n+1} \neq b_{n+1}$, we may suppose, after interchanging x and y, if necessary that $a_{n+1} = 1$ and $b_{n+1} = 2$. Let $x_0 = x, x_1, x_2, \ldots$ and $y_0 = y, y_1, y_2, \ldots$ be the numbers introduced in the construction of the continued fraction expansions of x and y. Then, as in (1),

$$x = \frac{1}{a_1+}\,\frac{1}{a_2+}\cdots\frac{1}{a_n+}\,\frac{1}{x_{n+1}},$$

$$y = \frac{1}{a_1+}\,\frac{1}{a_2+}\cdots\frac{1}{a_n+}\,\frac{1}{y_{n+1}}.$$

Further, as $a_{n+1} = 1$, we have

$$1 < x_{n+1} < 2.$$

Again, as $b_{n+1} = 2, b_{n+2} = 1$ or 2 and $y_{n+2} > 1$,

$$y_{n+1} = 2 + \frac{1}{y_{n+2}} < 3,$$

and
$$y_{n+1} = 2 + \frac{1}{b_{n+2}+}\,\frac{1}{y_{n+3}} > 2 + \frac{1}{2+1} = 2\tfrac{1}{3}.$$

Thus
$$1 < x_{n+1} < 2, \qquad 2\tfrac{1}{3} < y_{n+1} < 3.$$

So, using (7) of theorem 60,

$$\begin{aligned} |x-y| &= \frac{|x_{n+1}-y_{n+1}|}{(x_{n+1}q_n+q_{n-1})\,(y_{n+1}q_n+q_{n-1})} \\ &\geqslant \frac{\frac{1}{3}}{(2q_n+q_{n-1})\,(3q_n+q_{n-1})} \\ &\geqslant \frac{1}{27q_n(q_n+q_{n-1})} = \tfrac{1}{27}l(a_1, a_2, \ldots, a_n). \end{aligned}$$

As x and y both lie in the interval $I(a_1, a_2, \ldots, a_n)$ of length

$$l(a_1, a_2, \ldots, a_n)$$

the other half of the inequality (15) is immediate.

***Lemma* 3.** *The set P is closed.*

Proof. We first remark that the initial conditions $q_{-1} = 0$, $q_0 = 1$ and the recurrence relationship $q_{n+1} = a_{n+1}q_n + q_{n-1}$ together with the restriction $a_n \leqslant 2$, $n \geqslant 1$, immediately imply that

$$q_n \leqslant 3^n \quad (n \geqslant 0).$$

Hence we have

$$l(a_1, a_2, \ldots, a_n) = \frac{1}{q_n(q_n + q_{n-1})} \geqslant \tfrac{3}{4}3^{-2n}, \tag{16}$$

for $n \geqslant 1$.

Now suppose that $\{x^{(i)}\}$ is a convergent sequence of points of P. Let the continued fraction expansion of $x^{(i)}$ be

$$x^{(i)} = \frac{1}{a_1^{(i)}+}\,\frac{1}{a_2^{(i)}+}\ldots,$$

all the partial quotients $a_n^{(i)}$ taking only the values 1 or 2. Now, if n is a given positive integer, we can choose an integer $k(n)$ so large that

$$|x^{(i)} - x^{(j)}| < \tfrac{1}{36}3^{-2n},$$

for all i, j, with $k(n) \leqslant i < j$. Then, by (16), we have

$$|x^{(i)} - x^{(j)}| < \tfrac{1}{27} l(a_1^{(i)}, a_2^{(i)}, \ldots, a_r^{(i)}) \quad (r = 1, 2, \ldots, n).$$

It follows from lemma 2 that

$$a_1^{(i)} = a_1^{(j)}, \quad a_2^{(i)} = a_2^{(j)}, \ldots, a_n^{(i)} = a_n^{(j)}.$$

Thus, for a fixed positive integer r, $a_r^{(i)}$ takes a fixed value, a_r^* say, for all sufficiently large i.

Now the sequence $a_1^*, a_2^*, \ldots$ is a sequence of 1s and 2s and so there is an irrational number x^* in P with the expansion

$$x^* = \frac{1}{a_1^*+}\,\frac{1}{a_2^*+}\ldots.$$

But, for $i \geqslant k(n)$ we have

$$a_1^{(i)} = a_1^*, a_2^{(i)} = a_2^*, \ldots, a_n^{(i)} = a_n^*,$$

so that, by Lemma 2 and (11),

$$\begin{aligned} |x^{(i)} - x^*| &< l(a_1^*, a_2^*, \ldots, a_n^*) \\ &= \frac{1}{q_n^*(q_n^* + q_{n+1}^*)} \\ &< (\tfrac{1}{2} + \tfrac{1}{2}\sqrt{5})^{2-2n}. \end{aligned}$$

Hence the sequence $\{x^{(i)}\}$ converges to the point x^* of P. Thus P is closed.

2.3. These lemmas enable us to prove

***Theorem* 61.** *The set P of irrationals x, with $0 < x < 1$, that have continued fraction expansions with partial quotients not exceeding 2 is of positive $(\frac{1}{3})$-measure and of finite $(\frac{2}{3})$-measure. The same results hold* *for the set Q of irrationals x, with $0 < x < 1$, that have continued fraction expansions with at most a finite number of partial quotients exceeding* 2.

Proof. Let $\delta > 0$ be given. We can choose $n > 1$ so large that

$$(\tfrac{1}{2}+\tfrac{1}{2}\sqrt{5})^{2-2n} < \delta.$$

Then, by (11),
$$\frac{1}{q_n(q_n+q_{n-1})} < \delta,$$

and, by (10),
$$l(a_1, a_2, \ldots, a_n) < \delta,$$

for $a_1, a_2, \ldots, a_n = 1$ or 2. So P is covered by the intervals

$$I(a_1, a_2, \ldots, a_n), \qquad a_1, a_2, \ldots, a_n = 1 \text{ or } 2,$$

of length less than δ. Further, using lemma 1 repeatedly

$$\begin{aligned}
&\sum_{a_1, \ldots, a_n=1, 2} \{l(a_1, \ldots, a_n)\}^{\frac{2}{3}} \\
&\quad = \sum_{a_1, \ldots, a_{n-1}=1, 2} [\{l(a_1, a_{n-1}, 1)\}^{\frac{2}{3}} + \{l(a_1, \ldots, a_{n-1}, 2)\}^{\frac{2}{3}}] \\
&\quad \leqslant \sum_{a_1, \ldots, a_{n=1}=1, 2} \{l(a_1, \ldots, a_{n-1})\}^{\frac{2}{3}} \\
&\quad \leqslant \ldots \\
&\quad \leqslant \{l(1)\}^{\frac{2}{3}} + \{l(2)\}^{\frac{2}{3}} \leqslant 2.
\end{aligned}$$

Hence
$$\mu^{\frac{2}{3}}(P) \leqslant 2.$$

Now it is clear from the nature of the proof of lemma 1 that, if $\epsilon > 0$ is sufficiently small, we have

$$\{l(a_1, a_2, \ldots, a_n, 1)\}^{(\frac{2}{3}-\epsilon)} + \{l(a_1, a_2, \ldots, a_n, 2)\}^{(\frac{2}{3}-\epsilon)} \leqslant \{l(a_1, a_2, \ldots, a_n)\}^{(\frac{2}{3}-\epsilon)}.$$

Hence, it follows, by the above argument, that

$$\mu^{(\frac{2}{3}-\epsilon)}(P) \leqslant 2.$$

Consequently, by theorem 39,

$$\mu^{(\frac{2}{3})}(P) = 0.$$

Now Q is the union over all finite sequences $b_1, b_2, \ldots, b_n$ of positive integers of the sets $P(b_1, b_2, \ldots, b_n)$ of all irrational numbers

$$\frac{1}{b_1+}\,\frac{1}{b_2+}\cdots\frac{1}{b_n+}\,\frac{1}{a_1+}\,\frac{1}{a_2+}\cdots$$

with $a_1, a_2, \ldots$ equal to 1 or 2.

But $P(b_1, b_2, \ldots, b_n)$ is the image of P under the monotone map

$$x \to \frac{1}{b_1+}\,\frac{1}{b_2+}\cdots\frac{1}{b_{n-1}+}\,\frac{1}{b_n+x},$$

which has a bounded differential coefficient and an inverse with a bounded differential coefficient. Hence, by theorem 29,

$$\mu^{(\frac{2}{3})}(P(b_1, b_2, \ldots, b_n)) = 0.$$

Consequently $$\mu^{(\frac{2}{3})}(Q) = 0,$$
as required.

We now prove that $\mu^{(\frac{1}{3})}(P)$ is positive. As $P \subset Q$ this will imply that $\mu^{(\frac{1}{3})}(Q)$ is also positive. We need to study the coverings of P by intervals; but it will be enough to confine our attention to coverings of P by open intervals. Let $J_1, J_2, \ldots$ be a sequence of open intervals covering P. We aim to prove that

$$\sum_{i=1}^{\infty} \{d(J_i)\}^{\frac{1}{3}} \geqslant \tfrac{1}{3}. \tag{17}$$

As P is closed and bounded and the intervals are open, we can choose a positive integer M so that the finite sequence $J_1, J_2, \ldots, J_M$ covers P. Further, after some renaming, we may suppose that each of the intervals $J_1, J_2, \ldots, J_M$ meet P.

For each i with $1 \leqslant i \leqslant M$ we have $J_i \cap P \neq \varnothing$. Choose a point

$$x = \frac{1}{a_1+}\,\frac{1}{a_2+}\cdots$$

in $J_i \cap P$. As J_i is open, if we choose a sufficiently large integer n and consider the point

$$x' = \frac{1}{a_1+}\,\frac{1}{a_2+}\cdots\frac{1}{a_{n-1}+}\,\frac{1}{3-a_n+}\,\frac{1}{a_{n+1}+}\cdots,$$

obtained from x by replacing a_n by $3-a_n$ (to ensure that $3-a_n$ has the value 2 when $a_n = 1$ and the value 1 when $a_n = 2$) we obtain a second point in $J_i \cap P$. So far each i with $1 \leqslant i \leqslant M$ the set $J_i \cap P$ contains at least two points.

Now, for each i with $1 \leqslant i \leqslant M$, we can choose a unique integer $n(i) \geqslant 0$ so that the partial quotients

$$a_1, a_2, \ldots, a_{n(i)},$$

corresponding to the irrationals x in $J_i \cap P$ all agree, but there are two points of $J_i \cap P$ for which the partial quotient $a_{n(i)+1}$ takes different values. Suppose that the agreed values of the first $n(i)$ partial quotients for the points of $J_i \cap P$ are

$$a_1^{(i)}, a_2^{(i)}, \ldots, a_{n(i)}^{(i)}.$$

Then, by lemma 2, $\quad d(J_i) \geqslant \frac{1}{27} l(a_1^{(i)}, a_2^{(i)}, \ldots, a_{n(i)}^{(i)}).$

Further $$J_i \cap P \subset I(a_1^{(i)}, a_2^{(i)}, \ldots, a_{n(i)}^{(i)}). \tag{18}$$

Write $$I_i = I(a_1^{(i)}, a_2^{(i)}, \ldots, a_{n(i)}^{(i)}).$$

So $$d(J_i) \geqslant \tfrac{1}{27} l(a_1^{(i)}, a_2^{(i)}, \ldots, a_{n(i)}^{(i)}) = \tfrac{1}{27} d(I_i).$$

Thus $$\sum_{i=1}^{\infty} \{d(J_i)\}^{\frac{1}{3}} \geqslant \sum_{i=1}^{M} \{d(J_i)\}^{\frac{1}{3}} \geqslant \tfrac{1}{3} \sum_{i=1}^{M} \{d(I_i)\}^{\frac{1}{3}}. \tag{19}$$

Also by (18) we have $$P \subset \bigcup_{i=1}^{M} I_i. \tag{20}$$

So the inequality (17) will follow from (19), provided we can prove that

$$\sum_{i=1}^{M} \{d(I_i)\}^{\frac{1}{3}} \geqslant 1, \tag{21}$$

whenever $I_1, I_2, \ldots, I_M$ is a finite cover of P by intervals chosen from the family

$$I(a_1, a_2, \ldots, a_n) \quad (a_1, a_2, \ldots, a_n = 1 \quad \text{or} \quad 2; \quad n = 1, 2, \ldots).$$

It is clear that in proving this last result we may suppose that each interval meets P and that no interval is a sub-interval of any other interval of the given system $I_1, I_2, \ldots, I_M$.

We suppose that we have such a system $I_1, I_2, \ldots, I_M$. Suppose that

$$I_i = I(a_1^{(i)}, a_2^{(i)}, \ldots, a_{n(i)}^{(i)}),$$

for $1 \leqslant i \leqslant M$. Write

$$N = \max\{n(1), n(2), \ldots, n(M)\}.$$

Then one of the intervals is of the form

$$I_j = I(a_1^{(j)}, a_2^{(j)}, \ldots, a_{N-1}^{(j)}, a_N^{(j)}).$$

Now P contains points in the modified interval

$$I(a_1^{(j)}, a_2^{(j)}, \ldots, a_{N-1}^{(j)}, 3 - a_N^{(j)}).$$

If these points belonged to an interval

$$I_k = (a_1^{(k)}, a_2^{(k)}, \ldots, a_{n(k)}^{(k)}),$$

with $1 \leqslant k \leqslant M$ and $n(k) < N$, we would necessarily have

$$a_1^{(k)} = a_1^{(j)}, \quad a_2^{(k)} = a_2^{(j)}, \ldots, a_{n(k)}^{(k)} = a_{n(k)}^{(j)},$$

and we would have $$I_j \subset I_k,$$
contrary to our supposition. Hence the points, in the interval
$$I(a_1^{(j)}, a_2^{(j)}, \ldots, a_{N-1}^{(j)}, 3-a_N^{(j)}),$$
must be covered by precisely this interval, which must be one of the intervals I_i, $1 \leqslant i \leqslant M$, say the interval I_k. Thus both the intervals
$$\left.\begin{array}{l} I(a_1^{(j)}, a_2^{(j)}, \ldots, a_{N-1}^{(j)}, 1), \\ I(a_1^{(j)}, a_2^{(j)}, \ldots, a_{N-1}^{(j)}, 2), \end{array}\right\} \tag{22}$$
occur in the sequence $I_1, I_2, \ldots, I_M$. Let $I_1', I_2', \ldots, I_M'$ be the sequence obtained from the sequence $I_1, I_2, \ldots, I_M$ by replacing the two intervals (22) by the single interval
$$I(a_1^{(j)}, a_2^{(j)}, \ldots, a_{N-1}^{(j)}).$$
Then the intervals $I_1', I_2', \ldots, I_{M-1}'$ cover P. Also, by lemma 1,
$$\begin{aligned} \{d(I(a_1^{(j)}, \ldots, a_{N-1}^{(j)}, 1))\}^{\frac{1}{3}} &+ \{d(I(a_1^{(j)}, \ldots, a_{N-1}^{(j)}, 2))\}^{\frac{1}{3}} \\ &= \{l(a_1^{(j)}, \ldots, a_{N-1}^{(j)}, 1)\}^{\frac{1}{3}} + \{l(a_1^{(j)}, \ldots, a_{N-1}^{(j)}, 2)\}^{\frac{1}{3}} \\ &\geqslant \{l(a_1^{(j)}, \ldots, a_{N-1}^{(j)})\}^{\frac{1}{3}} \\ &= \{d(I(a_1^{(j)}, \ldots, a_{N-1}^{(j)}))\}^{\frac{1}{3}}, \end{aligned}$$
so that
$$\sum_{i=1}^{M} \{d(I_i)\}^{\frac{1}{3}} \geqslant \sum_{i=1}^{M-1} \{d(I_i')\}^{\frac{1}{3}}. \tag{23}$$

In this way we have replaced the cover of P by a similar cover of P with one less interval. By $M-2$ such steps we obtain such a cover by two intervals, the only possibility being the cover by $I(1)$ and $I(2)$. Using (23) at each step we obtain
$$\sum_{i=1}^{M} \{d(I_i)\}^{\frac{1}{3}} \geqslant \{l(1)\}^{\frac{1}{3}} + \{l(2)\}^{\frac{1}{3}} \geqslant 1,$$
the last inequality being a degenerate case of lemma 1. This completes the proof of the inequality (21). Hence
$$\mu^{(\frac{1}{3})}(P) \geqslant \tfrac{1}{3}.$$

§3 The space of non-decreasing continuous functions defined on the closed unit interval

In this section we give an account of a fraction of the work of S. J. Taylor and the author on additive set functions in Euclidean space (see C. A. Rogers and S. J. Taylor, 1963, and the references given there).

For simplicity we confine our attention to the unit interval of the real line and present the results as a contribution to the theory of non-decreasing continuous functions, defined on the closed unit interval $I_0 = [0, 1]$, and taking the value zero at the point zero. We use $\mathscr{P}$ to denote this space. As each function F of $\mathscr{P}$ is uniformly continuous on I, the function

$$h(t) = \sup_{\substack{0 \leqslant x < y \leqslant 1 \\ |x-y| \leqslant t}} |F(y) - F(x)|, \tag{1}$$

defined for $t \geqslant 0$, tends monotonically and continuously to 0 as t tends to 0 through positive values; and F satisfies the corresponding uniform Lipschitz condition

$$|F(y) - F(x)| \leqslant h(y-x),$$

for all x, y with $0 \leqslant x < y \leqslant 1$.

If a function F of $\mathscr{P}$ happens to be absolutely continuous, Lebesgue's theory shows that

$$f(t) = F'(t),$$

is defined for almost all t in I_0, and that

$$F(t) = \int_0^t f(s)\, ds,$$

for $0 \leqslant t \leqslant 1$. Our aim will be to explore the possibility of extending this result in some way to all the functions of $\mathscr{P}$; but we shall see that the extension will need various new ideas.

We shall first need to introduce some continuity concepts intermediate between continuity and absolute continuity. We recall an indirect approach to the theory of absolute continuity. Starting from any function F of $\mathscr{P}$, we can form a pre-measure τ_F, by defining

$$\tau_F([a, b]) = F(b) - F(a),$$

for each closed sub-interval $[a, b]$ of I_0, and, as always, taking

$$\tau_F(\varnothing) = 0;$$

and we can then introduce the measure constructed from τ_F by Method I. We use F itself to denote this measure constructed from the function F of P; this should cause little confusion. In theorem 62 below, we show that F is also the measure constructed from F by Method II and that $F([a, b]) = F(b) - F(a)$, for all closed sub-intervals of I_0. We adopt the following definition of the absolute continuity of the function F in terms of the properties of the measure F.

***Definition* 33.** *The function F of $\mathscr{P}$ is said to be absolutely continuous, if*

$$F(E) = 0,$$

whenever E is a subset of I_0 of Lebesgue measure zero.

By analogy with this definition we introduce two concepts of continuity associated with each function h of $\mathscr{H}_0$.

***Definition* 34.** *A function F of $\mathscr{P}$ is said to be h-continuous, if*

$$F(E) = 0,$$

whenever E is a subset of I_0 of μ^h-measure zero.

***Definition* 35.** *A function F of $\mathscr{P}$ is said to be strongly h-continuous, if*

$$F(E) = 0,$$

whenever E is a subset of I_0 of finite μ^h-measure.

We will relate these concepts of continuity to the Lipschitz conditions associated with the function h of $\mathscr{H}_0$, proving for example that a function F of $\mathscr{P}$ is h-continuous, if, and only if, it can be expressed as the sum of functions $\{G_i\}$ of P, each G_i satisfying a Lipschitz condition (see theorem 68 below). The Lipschitz conditions that we need are those of the following definitions.

***Definition* 36.** *If $h \in \mathscr{H}_0$, a function F of $\mathscr{P}$ will be said to be uniformly Lip-h if there is a finite number K such that*

$$F(b) - F(a) \leqslant Kh(b-a),$$

for all a, b with $0 \leqslant a < b \leqslant 1$.

***Definition* 37.** *If $h \in \mathscr{H}_0$, a function F of $\mathscr{P}$ will be said to be uniformly lip-h if for each $\kappa > 0$ there is a $\delta > 0$ such that*

$$F(b) - F(a) \leqslant \kappa h(b-a),$$

for all a, b with $0 \leqslant a < b \leqslant 1$ and $b - a < \delta$.

Here we use *Lip-h* and *lip-h* in analogy with the Landau O-, o-notation.

Our aim (see theorem 69 below) will be to show that if $h \in \mathscr{H}_0$ and F in $\mathscr{P}$ is h-continuous then, for all Borel sets E contained in I_0,

$$F(E) = \int_E f(t)\, d\mu^h + J(E),$$

where f is a point function on I_0 and J is a function of $\mathscr{P}$ that is strongly h-continuous. Although we will be able to give detailed proofs, this last result will depend on the Radon–Nikodym theorem, a result that we shall not prove.

3.1. After this general discussion we proceed to the statement and proof of

***Theorem* 62.** *Let F belong to $\mathscr{P}$ and let τ_F be the pre-measure defined on the closed sub-intervals of I_0 by the formula*

$$\tau_F([a,b]) = F(b)-F(a).$$

The measures constructed from τ_F by Methods I and II coincide and assign the measures $F(b)-F(a)$ to the closed sub-intervals $[a,b]$ of I_0.

Proof. We use F to denote the Method I measure constructed from τ_F, and F_{II} to denote the Method II measure constructed from τ_F. Then, for all E in I,

$$F(E) \leqslant F_{\text{II}}(E).$$

To prove that

$$F_{\text{II}}(E) \leqslant F(E),$$

it suffices to consider the case when $F(E)$ is finite. Then, if $\delta > 0$ is given and $\epsilon > 0$ is given, we can choose a sequence $\{[a_i, b_i]\}$ of closed sub-intervals of I_0 with

$$E \subset \bigcup_{i=1}^{\infty} [a_i, b_i], \qquad \sum_{i=1}^{\infty} \tau_F([a_i, b_i]) < F(E)+\epsilon.$$

Now, for each i, the interval $[a_i, b_i]$ is the union of a finite sequence

$$[a_{ir}, b_{ir}] \quad (r = 1, 2, \ldots, N(i)),$$

say, of adjacent non-overlapping intervals each of length less than δ. So

$$\sum_{r=1}^{N(i)} \tau_F([a_{ir}, b_{ir}]) = \tau_F([a_i, b_i]),$$

and

$$E \subset \bigcup_{i=1}^{\infty} \bigcup_{r=1}^{N(i)} [a_{ir}, b_{ir}],$$

while

$$\sum_{i=1}^{\infty} \sum_{r=1}^{N(i)} \tau_F([a_{ir}, b_{ir}]) = \sum_{i=1}^{\infty} \tau_F([a_i, b_i]) < F(E)+\epsilon.$$

Consequently $F_{\text{II}}\ (E) < F(E)+\epsilon$, for each $\epsilon > 0$. Thus we obtain $F_{\text{II}}(E) = F(E)$ as required.

Now consider any closed sub-interval $[a,b]$ of I_0. Clearly

$$F([a,b]) \leqslant \tau_F([a,b]) = F(b)-F(a).$$

Suppose we had

$$F([a,b]) < F(b)-F(a).$$

Then we could choose a sequence $\{[a_i, b_i]\}$ of closed sub-intervals of I_0 with

$$[a, b] \subset \bigcup_{i=1}^{\infty} [a_i, b_i],$$

$$\sum_{i=1}^{\infty} \tau_F([a_i, b_i]) < F(b) - F(a). \tag{2}$$

Thus
$$\sum_{i=1}^{\infty} \{F(b_i) - F(a_i)\} < F(b) - F(a).$$

As F is continuous and increasing in I_0, the sequence of intervals $\{[F(a_i), F(b_i)]\}$ covers the closed interval $[F(a), F(b)]$. But, by (2), the sum of the lengths of the intervals of the covering sequence is less than the length of the interval covered. This contradicts a basic property of Lebesgue measure (theorem 25). Consequently we must have

$$F([a, b]) = F(b) - F(a),$$

as required.

Remark. The first part of this proof is essentially the same as the proof of theorem 24. We have reduced the proof of the second part to an application of theorem 25; it could equally well have been proved using the method used to prove theorem 25. Radon (1913) working in an n-dimensional interval introduced such measures $F(E)$ associated with point functions $F(x_1, x_2, \ldots, x_n)$ that are increasing in a generalized sense.

3.2. We use the following definition for an upper h-derivative of a function F of $\mathscr{P}$, and then use it as a tool to study the sets where the function F is 'varying more or less rapidly'.

***Definition* 38.** *If h belongs to $\mathscr{H}_0$ and F belongs to $\mathscr{P}$ and $x \in I_0$, the upper h-derivative $\bar{D}_h F(x)$ of F at x is defined by*

$$\bar{D}_h F(x) = \lim_{\delta \to 0} \sup_{\substack{x \in I \\ d(I) < \delta}} \frac{F(I)}{h(I)}, \tag{3}$$

the supremum being taken over the set of all sub-intervals I of I_0, open relative to I_0, containing x and having length less than δ.

We first use a Vitali argument to show that the set of points, where the upper h-derivative is large, has small μ^h-measure. More precisely, we prove

***Theorem* 63.** *Suppose that $h \in \mathscr{H}_0$ and that $F \in \mathscr{P}$. For each positive real number k, let U_k be the set of points x of I_0 with*

$$\bar{D}_h F(x) > k. \tag{4}$$

Then U_k is a $\mathscr{G}_{\delta\sigma}$-set and $$\mu^h(U_k) \leqslant \frac{5}{k} F(I_0). \tag{5}$$

Proof. For all pairs m, n of positive integers let $G_{m,n}$ be the set of all x in I_0 for which

$$\sup_{\substack{x \in I \\ d(I)<1/n}} \frac{F(I)}{h(I)} > k + \frac{1}{m}, \tag{6}$$

the supremum being taken over all the sub-intervals I of I_0, open relative to I_0, containing x and having length less than $1/n$. Consider any point x^* of $G_{m,n}$. It follows, from (6), that there is an interval I^* contained in I_0, open relative to I_0, containing x^*, and having length less than $1/n$ and satisfying

$$\frac{F(I^*)}{h(I)^*} > k + \frac{1}{m}.$$

Hence for each point x of I^* we have

$$\sup_{\substack{x \in I \\ d(I)<1/n}} \frac{F(I)}{h(I)} \geqslant \frac{F(I^*)}{h(I^*)} > k + \frac{1}{m},$$

and so I^* is contained in $G_{n,m}$. As $x^* \in I^*$ and I^* is open relative to I_0, it follows that $G_{n,m}$ is open relative to I_0.

Now the supremum

$$\sup_{\substack{x \in I \\ d(I)<\delta}} \frac{F(I)}{h(I)}$$

does not increase when δ is decreased. Hence, for x in I_0,

$$\begin{aligned} \bar{D}_h F(x) &= \lim_{\delta \to 0} \sup_{\substack{x \in I \\ d(I)<\delta}} \frac{F(I)}{h(I)} \\ &= \lim_{n \to \infty} \sup_{\substack{x \in I \\ d(I)<1/n}} \frac{F(I)}{h(I)}. \end{aligned}$$

So the condition on x in I_0 that

$$\bar{D}_h F(x) > k,$$

is equivalent to the condition that

$$\lim_{n \to \infty} \sup_{\substack{x \in I \\ d(I)<1/n}} \frac{F(I)}{h(I)} > k,$$

and so to the condition that, for some positive integer m and all positive integers n,

$$\sup_{\substack{x \in I \\ d(I)<1/n}} \frac{F(I)}{h(I)} > k+\frac{1}{m}.$$

Hence
$$U_k = \bigcup_{m=1}^{\infty} \bigcap_{n=1}^{\infty} G_{n,m},$$

and U_k is a $\mathcal{G}_{\delta\sigma}$-set.

Now to estimate the μ^h-measure of U_k we need to find an economical fine cover of U_k. Let $\delta > 0$ be given. If we could cover U_k by a disjoint sequence $J_1, J_2, \ldots$ of intervals, with $d(J_i) < \delta, i = 1, 2, \ldots$, and with
$$F(J_i) \geqslant kh(J_i) \quad (i = 1, 2, \ldots),$$

we could argue that the total variation

$$\sum_{i=1}^{\infty} F(J_i)$$

of F over the intervals did not exceed the total variation $F(I_0)$ of F over I_0, and we could deduce that

$$\sum_{i=1}^{\infty} h(J_i) \leqslant \frac{1}{k} F(I_0), \tag{7}$$

and conclude that
$$\mu^h(U_k) \leqslant \frac{1}{k} G(I_0). \tag{8}$$

Although the points of U_k are necessarily included in arbitrarily small intervals J with $F(J) > kh(J)$, it is not a straightforward matter to find a disjoint system of such intervals covering U_k. So we use a most ingenious argument, due to Vitali, to select from the intervals J with $F(J) > kh(J)$ a disjoint sequence $J_1, J_2, \ldots$, necessarily satisfying (7), so that this system of intervals, augmented by a system of 'flanking' intervals (two on each side of each interval of the sequence $J_1, J_2, \ldots$) covers U_k. Although this does not lead to a proof of (8) it does lead to the required result (5).

For each x in U_k we choose an interval $I(x)$, open relative to I_0, with x in $I(x)$, with $d(I(x)) < \delta$, and with

$$F(I(x)) > kh(I(x)).$$

Let $J(x)$ be the closure of $I(x)$. Then, clearly,

$$F(J(x)) > kh(J(x)). \tag{9}$$

It is convenient to use $\mathcal{J}$ to denote the family of all these intervals $J(x)$.

The intervals J of $\mathscr{J}$ certainly cover U_k; but in general, there will be far too many of them. We proceed to pick out a disjoint subsystem of these intervals. Amongst the intervals J of $\mathscr{J}$ there may or may not be a largest one. But certainly

$$\sup_{J \in \mathscr{J}} d(J) \leqslant d(I_0) = 1.$$

So we can choose J_1 in $\mathscr{J}$ with

$$d(J_1) > \tfrac{1}{2} \sup_{J \in \mathscr{J}} d(J).$$

We choose a finite or infinite sequence $J_1, J_2, \ldots$ inductively. When a disjoint sequence $J_1, J_2, \ldots, J_n$ of intervals of $\mathscr{J}$ have been chosen, it may or may not be possible to choose a further interval J of $\mathscr{J}$ disjoint from $J_1, J_2, \ldots,, J_n$. If there is no such J in $\mathscr{J}$ we terminate our sequence. If there is such a J in $\mathscr{J}$ we choose one of the larger such intervals; more precisely we choose J_{n+1} in $\mathscr{J}$ so that

$$J_1, J_2, \ldots, J_n, J_{n+1}$$

are disjoint, and $$d(J_{n+1}) > \tfrac{1}{2} \sup_{\substack{J \in \mathscr{J} \\ (J_1 \cup J_2 \cup \ldots \cup J_n) \cap J = \emptyset}} d(J). \tag{10}$$

We obtain in this way a finite or infinite sequence of intervals

$$J_1, J_2, \ldots (J_m)$$

of $\mathscr{J}$, all mutually disjoint; here we place the J_m in brackets to remind us of the possibility that the sequence terminates with J_m.

We write

$$J_{i0} = J_i \quad (i = 1, 2, \ldots, (m)),$$

and we flank J_{i0} by equal intervals $J_{i(-2)}, J_{i(-1)}, J_{i1}, J_{i2}$, two on each side. These flanking intervals take the form

$$[a-2h, a-h], [a-h, a], [a+h, a+2h], [a+2h, a+3h],$$

when $J_{i0} = J_i$ has the form $\quad [a, a+h];$

they may well include points outside I_0.

Our first aim will be to show that

$$\bigcup_i \bigcup_{j=-2}^{2} J_{ij} \supset U_k.$$

Suppose that $x \in U_k$. Then $J(x) \in \mathscr{J}$. Suppose that $J(x)$ meets none of the intervals $J_1, J_2, \ldots, (J_m)$. If the sequence should terminate with the interval J_m, then the inductive construction is such that $J(x)$ is an *a priori* possible choice for J_{m+1}, and either $J(x)$ should have been taken for J_{m+1} or some longer interval should have been taken. So we suppose the sequence $J_1, J_2, \ldots$ to be infinite. As the intervals are disjoint sub-intervals of I_0, this implies that $d(J_i) \to 0$ as $i \to \infty$. So $d(J_{n+1}) < \frac{1}{2}d(J(x))$, for some sufficiently large integer n. But now, as

$$J_1, J_2, \ldots, J_n, J(x)$$

are disjoint, we have

$$d(J_{n+1}) < \tfrac{1}{2}d(J(x)) \leqslant \tfrac{1}{2} \sup_{\substack{J \in \mathscr{J} \\ (J_1 \cup J_2 \cup \ldots \cup J_n) \cap J = \varnothing}} d(J),$$

contrary to the choice of J_{n+1} by our inductive procedure. This shows that $J(x)$ meets at least one of the intervals $J_1, J_2, \ldots (J_m)$.

Let $J_{n(x)}$ be the first of the intervals $J_1, J_2, \ldots (J_m)$ that meets $J(x)$. Then $J(x)$ does not meet $J_1, J_2, \ldots, J_{n(x)-1}$, and so is a candidate available as a possible choice for $J_{n(x)}$. So $J(x)$ cannot be twice as long as the interval $J_{n(x)}$ that was chosen at this stage. Thus

$$d(J(x)) \leqslant 2d(J_{n(x)}).$$

As $J(x)$ meets $J_{n(x)}$, it follows that $J(x)$, and so x, is contained in

$$\bigcup_{j=-2}^{2} J_{n(x)j}$$

As x was any point of U_k, it follows that

$$U_k \subset \bigcup_i \bigcup_{j=-2}^{2} J_{ij}.$$

Now

$$\sum_i \sum_{j=-2}^{2} h(J_{ij}) = 5 \sum_i h(J_i) \leqslant (5/k) \sum_i F(J_i) = (5/k)\, F(\bigcup_i J_i) \leqslant (5/k)\, F(I_0),$$

using (9), the disjointness of the sequence $J_1, J_2, \ldots (J_m)$, and the properties of the measure F. As the system of sets J_{ij}, $i = 1, 2, \ldots (m)$, $j = -2, -1, 0, 1, 2$, is a cover of U_k by sets of diameter less than δ, we conclude that

$$\mu^h(U_k) \leqslant (5/k)\, F(I_0),$$

as required.

It is now relatively easy to obtain corresponding results for the set U_∞ of points x of I where the upper h-derivative is infinite.

Theorem 64. *Suppose that $h \in \mathscr{H}_0$ and that $F \in \mathscr{P}$. Let U_∞ be the set of points x of I_0 with*

$$\overline{D}_h F(x) = +\infty. \tag{11}$$

Then U_∞ is a $\mathscr{G}_{\delta\sigma\delta}$-set, and

$$\mu^h(U_\infty) = 0. \tag{12}$$

Proof. With the notation of theorem 63.

$$U_\infty = \bigcap_{k=1}^{\infty} U_k. \tag{13}$$

It follows, from theorem 63, that U_∞ is a $\mathscr{G}_{\delta\sigma\delta}$-set.

Also, by (13) and (5) of theorem 63, for each positive integer k, we have

$$\mu^h(U_\infty) \leqslant \mu^h(U_k) \leqslant (5/k)\, F(I_0).$$

This shows that (12) must hold.

3.3. Our next aim is to show that the F-measures of those subsets of small μ^h-measure of the set where the upper h-derivative is small are themselves small. We first prove

Theorem 65. *Suppose that $h \in \mathscr{H}_0$ and that $F \in \mathscr{P}$. For each positive real number k and each δ with $0 < \delta \leqslant 1$, let $L_{k,\delta}$ be the set of points x of I_0 with*

$$\sup_{\substack{x \in I \\ d(I) \leqslant \delta}} \frac{F(I)}{h(I)} \leqslant k, \tag{14}$$

the supremum being taken over all the sub-intervals I of I_0 that contain x and are open relative to I_0. Then $L_{k,\delta}$ is a closed set and the function

$$G_{k,\delta}(t) = F(L_{k,\delta} \cap [0,t]) \tag{15}$$

is in $\mathscr{P}$ and satisfies the Lipschitz condition

$$G_{k,\delta}(I) \leqslant kh(I), \tag{16}$$

for all sub-intervals I of I_0 with $d(I) \leqslant \delta$. Further

$$F(E \cap L_{k,\delta}) \leqslant k\mu^h(E), \tag{17}$$

for all subsets E of I_0.

Proof. Clearly $L_{k,\delta}$ is the complement with respect to I_0 of the set of points x of I_0 with

$$\sup_{\substack{x \in I \\ d(I) \leqslant \delta}} \frac{F(I)}{h(I)} > k.$$

But, just as in the discussion of the set $G_{m,n}$, in the first paragraph of the proof of theorem 63, this set is open relative to I_0. Hence $L_{k,\delta}$ is closed.

As F is a measure, and Borel sets are F-measurable, it is clear that $G_{k,\delta}(t)$ is monotonic increasing for $0 \leqslant t \leqslant 1$. Further $G_{k,\delta}(0) = 0$, and if $0 \leqslant a < b \leqslant 1$,

$$G_{k,\delta}(b) - G_{k,\delta}(a) = F(L_{k,\delta} \cap [a,b]) \leqslant F([a,b]) = F(b) - F(a),$$

so that $\mathscr{G}_{k,\delta}$ is continuous on I_0. Thus $G_{k,\delta} \in \mathscr{P}$.

As $\mathscr{G}_{k,\delta} \in \mathscr{P}$ we can use $G_{k,\delta}$ to denote the associated measure, as well as the function defined on I; indeed we have already anticipated this in formula (16).

Now consider sub-interval I of I_0 that is open relative to I_0 and has $d(I) \leqslant \delta$. If $I \cap L_{k,\delta}$ is empty, we have

$$G_{k,\delta}(I) = F(L_{k,\delta} \cap I) = F(\varnothing) = 0 \leqslant kh(I).$$

On the other hand, if $I \cap L_{k,\delta}$ is non-empty, we can choose a point x in $I \cap L_{k,\delta}$. Then $x \in L_{k,\delta}$, $x \in I$, and $d(I) \leqslant \delta$, and so, by the definition of $L_{k,\delta}$,

$$F(I) \leqslant kh(I).$$

In this case, we have $\quad G_{k,\delta}(I) \leqslant F(I) \leqslant kh(I).$

Thus (16) always holds for I open relative to I_0 with $d(I) \leqslant \delta$. Consequently (16) holds for all sub-intervals I of I_0, with $d(I) \leqslant \delta$.

Now consider any subset E of I_0. If $\mu^h(E)$ is infinite (17) is trivial. Suppose, then, that $\mu^h(E)$ is finite. For each $\epsilon > 0$, we can cover E by a sequence $\{J_i\}$ of sub-intervals of I_0, with

$$d(J_i) \leqslant \delta, \qquad \sum_{i=1}^{\infty} h(J_i) < \mu^h(E) + \epsilon.$$

Thus we have

$$\begin{aligned} F(E \cap L_{k,\delta}) &\leqslant F\left(\bigcup_{i=1}^{\infty} \{L_{k,\delta} \cap J_i\}\right) \\ &\leqslant \sum_{i=1}^{\infty} F(L_{k,\delta} \cap J_i) \\ &= \sum_{i=1}^{\infty} G_{k,\delta}(J_i) \\ &\leqslant k \sum_{i=1}^{\infty} h(J_i) \\ &\leqslant k\{\mu^h(E) + \epsilon\}. \end{aligned}$$

So (17) follows, and the proof is complete.

Corollary. *For all subsets E of I_0, we have*

$$F(E \cap L_{k,\delta}) \leqslant k\mu_\delta^h(E). \tag{18}$$

Proof. The proof of the last paragraph applies with μ_δ^h written for μ^h throughout (the case, when $\mu_\delta^h(E)$ is infinite, does not arise and can be omitted).

We can now easily establish

***Theorem* 66.** *Suppose that $h \in \mathscr{H}_0$ and that $F \in \mathscr{P}$. For each positive real number k, let L_k be the set of points x of I_0 with*

$$\bar{D}_h F(x) < k. \tag{19}$$

Then L_k is an $\mathscr{F}_\sigma$-set and

$$F(E \cap K_k) \leqslant k\mu^h(E), \tag{20}$$

for all subsets E of I_0.

Further, if T_0 is the set of all points x with $\bar{D}_h F(x) = 0$, then T_0 is an $\mathscr{F}_{\sigma\delta}$-set, and

$$F(E \cap T_0) = 0,$$

whenever E has σ-finite μ^h measure.

Proof. As the supremum

$$\sup_{\substack{x \in I \\ d(I) \leqslant \delta}} \frac{F(I)}{h(I)}$$

does not increase when δ decreases, we have

$$\lim_{\delta \to 0} \sup_{\substack{x \in I \\ d(I) \leqslant \delta}} \frac{F(I)}{h(I)} < k$$

if, and only if, there is a positive integer n for which

$$\sup_{\substack{x \in I \\ d(I) \leqslant 1/n}} \frac{F(I)}{h(I)} \leqslant k - (1/n).$$

Thus, using the notation of theorem 65, we have

$$L_k = \bigcup_{n=1}^{\infty} L_{k-(1/n),(1/n)}.$$

Hence, using theorem 65, the set L_k is an $\mathscr{F}_\sigma$-set.

Now, as the sets $L_{k-(1/n),(1/n)}$, $n = 1, 2, \ldots,$ are F-measurable, it follows, from theorem 8, that, for any set E,

$$\begin{aligned} F(E \cap L_k) &= F\left(E \cap \bigcup_{n=1}^{\infty} L_{k-(1/n),(1/n)}\right) \\ &\leqslant F\left(E \bigcup_{n=1}^{\infty} L_{k,(1/n)}\right) \\ &\leqslant \sup_n F(E \cap L_{k,(1/n)}) \\ &\leqslant k\mu^h(E), \end{aligned}$$

on using (17) of theorem 65. This proves (20).

Since
$$T_0 = \bigcap_{n=1}^{\infty} L_{(1/n)},$$
the set T_0 is a $\mathscr{F}_{\sigma\delta}$-set. Further, for each positive integer n,
$$F(E \cap T_0) \leqslant F(E \cap L_{(1/n)}) \leqslant (1/n)\,\mu^h(E).$$
Hence
$$F(E \cap T_0) = 0, \tag{21}$$
whenever E is of finite μ^h-measure. But, if $E = \bigcup_{i=1}^{i=\infty} E_i$ with each E_i of finite μ^h-measure, we deduce that
$$F(E \cap T_0) \leqslant \sum_{i=1}^{\infty} F(E_i \cap T_0) = 0.$$
Thus (21) holds whenever E is of σ-finite μ^h-measure.

3.4. It is convenient to summarize the main results of §§ 3.2, 3.3 in the following:

***Theorem* 67.** *Suppose that $h \in \mathscr{H}_0$ and that $F \in \mathscr{P}$. Let T_0, T_+, T_∞ be the sets of points x of I_0 where $\bar{D}_h F(x)$ takes the value 0, takes a finite positive value, or takes the value $+\infty$, respectively. Then T_0, T_+ and T_∞ are Borel sets;*

(*a*) $\mu^h(T_\infty) = 0$;
(*b*) *T_+ is of σ-finite μ^h-measure;*
(*c*) *$F(E \cap T_+) = 0$, if E has zero μ^h-measure; and*
(*d*) *$F(E \cap T_0) = 0$, if E is of σ-finite μ^h-measure.*

Proof. We have
$$T_\infty = U_\infty,$$
$$T_+ = \left\{\bigcup_{n=1}^{\infty} U_{(1/n)}\right\} \cap \left\{\bigcup_{n=1}^{\infty} L_n\right\},$$
$$T_0 = \bigcap_{n=1}^{\infty} L_{(1/n)}.$$
Thus the three sets T_0, T_+, T_∞ are Borel sets, by theorems 63 and 66. Further (*a*) follows from (12) of theorem 64; (*b*) follows from (5) of theorem 63; (*c*) follows from (20) of theorem 66; and (*d*) is (21) of theorem 66.

3.5. We now proceed to characterisations of the h-continuous and strongly h-continuous functions of $\mathscr{P}$ as sums of functions satisfying Lipschitz conditions.

Theorem 68. *Suppose that $h \in \mathscr{H}_0$ and that $F \in \mathscr{P}$. Then F is h-continuous if, and only if, F has a convergent representation*

$$F(t) = \sum_{i=1}^{\infty} G_i(t) \quad (0 \leqslant t \leqslant 1),$$

the functions G_i, $i = 1, 2, \ldots$, belonging to $\mathscr{P}$ and being uniformly Lip-h. Further F is strongly h-continuous if, and only if, F has a similar representation as a sum of uniformly lip-h functions of $\mathscr{P}$.

Proof. (*a*) Suppose that the function F of $\mathscr{P}$ has a convergent representation

$$F(t) = \sum_{i=1}^{\infty} G_i(t) \quad (0 \leqslant t \leqslant 1),$$

the functions G_i, $i = 1, 2, \ldots$, belonging to $\mathscr{P}$ and being uniformly *Lip*-h. To prove that F is h-continuous we study any set E with $\mu^h(E) = 0$ and seek to deduce that $F(E) = 0$. Let E be such a set, and let $\epsilon > 0$ be given. As $\Sigma\, G_i(1)$ converges, we can choose N so large that

$$\sum_{i=N+1}^{\infty} G_i(1) < \tfrac{1}{2}\epsilon. \tag{22}$$

For $i = 1, 2, \ldots, N$, we can choose constants K_i so that

$$G_i(I) \leqslant K_i h(I), \tag{23}$$

for all sub-intervals I of I_0. Now, as $\mu^h(E) = 0$, we cover E by a sequence $\{J_j\}$ of closed sub-intervals of I_0, with

$$\sum_{j=1}^{\infty} h(J_j) < \tfrac{1}{2}\epsilon \Big/ \left(\sum_{i=1}^{N} K_i\right). \tag{24}$$

By the convergence of the sum $\Sigma h(J_j)$ it is clear that there cannot be an infinite subsequence $J_j(k)$, $k = 1, 2, \ldots$ with

$$J_{j(1)} \subset J_{j(2)} \subset \ldots.$$

Hence each member of the sequence $\{J_j\}$ is contained in some member of the sequence that is maximal in that it is not properly contained in any member of the sequence. Replacing the sequence $\{J_j\}$ by its maximal members, we obtain a sequence $\{J_j^*\}$, that may be finite, covering E and satisfying

$$\Sigma\, h(J_j^*) < \tfrac{1}{2}\epsilon \Big/ \left(\sum_{i=1}^{N} K_i\right).$$

We suppose the sequence augmented by a sequence of copies of $\varnothing$, when it is finite, and we drop the asterisks from the notation.

We modify the sequence $\{J_j\}$ further, by replacing each interval J_j by a sub-interval† J_j^* that does not meet any of the intervals $J_1^*, J_2^*, \ldots, J_{j-1}^*$, except perhaps at a single point. We start by taking $J_1^* = J_1$. When $J_1^*, J_2^*, \ldots, J_{j-1}^*$ have been chosen, J_j is contained in none of these intervals and contains none of them. Thus, if one end-point of one of the intervals $J_1^*, J_2^*, \ldots, J_{j-1}^*$ lies in the interior of J_j, the second end-point must lie outside J_j, either to the left or to the right of J_j. It follows that J_j can meet at most two of the intervals $J_1^*, J_2^*, \ldots, J_{j-1}^*$. Hence the closure of

$$J_j \backslash \{J_1^* \cup J_2^* \cup \ldots \cup J_{j-1}^*\}$$

is a closed sub-interval of J_j, or reduces to $\varnothing$, when J_j is covered by two abutting intervals from among $J_1^*, J_2^*, \ldots, J_{j-1}^*$. We take J_j^* to be this closure. Proceeding inductively in this way we obtain a sequence $\{J_j^*\}$ of sub-intervals of I_0 with

$$J_j^* \subset J_j, \qquad \bigcup_{j=1}^{\infty} J_j^* = \bigcup_{j=1}^{\infty} J_j,$$

and

$$J_j^* \cap J_k^*$$

consisting of at most one point when $j \neq k$. These modified intervals still cover E and satisfy

$$\sum_{j=1}^{\infty} h(J_j^*) < \tfrac{1}{2}\epsilon \Big/ \left(\sum_{i=1}^{N} K_i\right). \tag{25}$$

Let J_j^* be the interval $[a_j^*, b_j^*]$ or $\varnothing$. Let $J_j^{(0)}$ denote the interior of J_j^*. Then using Σ' to denote a summation over the set of j with $J_j^* \neq \varnothing$,

$$\begin{aligned}
\sum_{j=1}^{\infty} F(J_j^*) &= \Sigma'\{F(b_j^*) - F(a_j^*)\} \\
&= \Sigma'\left\{\sum_{i=1}^{\infty} G_i(b_j^*) - \sum_{i=1}^{\infty} G_i(a_i^*)\right\} \\
&= \sum_{i=1}^{N} \Sigma' G_i(J_j^*) + \sum_{i=N+1}^{\infty} \Sigma' G_i(J_j^{(0)}) \\
&\leqslant \sum_{i=1}^{N} \Sigma' K_i h(J_j^*) + \sum_{i=N+1}^{\infty} G_i(I_0) \\
&\leqslant \left(\sum_{i=1}^{N} K_i\right)\left(\sum_{j=1}^{\infty} h(J_j^*)\right) + \sum_{i=N+1}^{\infty} G_i(1) \\
&\leqslant \tfrac{1}{2}\epsilon + \tfrac{1}{2}\epsilon = \epsilon,
\end{aligned}$$

† Here we allow $\varnothing$ to be a 'sub-interval'.

on using (23), (25) and (22). As $\epsilon > 0$ is arbitrary it follows that $F(E) = 0$. Thus F is h-continuous.

(*b*) Now suppose that the function F of $\mathscr{P}$ has a convergent representation

$$F(t) = \sum_{i=1}^{\infty} G_i(t) \quad (0 \leqslant t \leqslant 1),$$

the functions G_i, $i = 1, 2, \ldots$, belonging to $\mathscr{P}$ and being uniformly *lip-h*. To prove that F is strongly h-continuous we suppose that E is any set with finite μ^h-measure and seek to deduce that $F(E) = 0$. Let E be such a set, and let $\epsilon > 0$ be given. We modify the argument of part (*a*) slightly. Just as in part (*a*) we choose N so large that

$$\sum_{i=N+1}^{\infty} G_i(1) < \tfrac{1}{2}\epsilon. \tag{26}$$

We then choose positive constants $\kappa_1, \kappa_2, \ldots, \kappa_N$ with

$$\left(\sum_{i=1}^{N} \kappa_i\right)(1 + \mu^h(E)) < \tfrac{1}{2}\epsilon. \tag{27}$$

As the functions $G_1, G_2, \ldots, G_N$ are uniformly *lip-h*, we can choose $\delta > 0$ so that

$$G_i(I) \leqslant \kappa_i h(I), \tag{28}$$

for $i = 1, 2, \ldots, N$, for all I contained in I_0 with $d(I) < \delta$.

We cover E by a sequence $\{J_j\}$ of closed sub-intervals of I_0, with

$$d(J_j) < \delta, \qquad \sum_{j=1}^{\infty} h(J_j) < \mu^h(E) + 1.$$

By the process of modification described in part (*a*), we may replace the sequence $\{J_j\}$ by a sequence $\{J_j^*\}$ of closed sub-intervals (perhaps empty) of I_0, covering E with disjoint interiors $\{J_j^{(0)}\}$, and satisfying

$$d(J_j^*) < \delta, \qquad \sum_{j=1}^{\infty} h(J_j^*) < \mu^h(E) + 1. \tag{29}$$

Then, just as in part (*a*),

$$\begin{aligned}
\sum_{j=1}^{\infty} F(J_j^*) &= \sum_{i=1}^{N} \Sigma' G_i(J_j^*) + \sum_{i=N+1}^{\infty} \Sigma' G_i(J_j^{(0)}) \\
&\leqslant \sum_{i=1}^{N} \Sigma' \kappa_i h(J_j^*) + \sum_{i=N+1}^{\infty} G_i(I_0) \\
&\leqslant \left(\sum_{i=1}^{N} \kappa_i\right)\left(\sum_{j=1}^{\infty} h(J_j^*)\right) + \sum_{i=N+1}^{\infty} G_i(1) \\
&\leqslant \left(\sum_{i=1}^{N} \kappa_i\right)(1 + \mu^h(E)) + \tfrac{1}{2}\epsilon \\
&\leqslant \tfrac{1}{2}\epsilon + \tfrac{1}{2}\epsilon = \epsilon,
\end{aligned}$$

on using (28), (29), (26) and (27). Thus $F(E) = 0$ and F is strongly h-continuous.

(c) Suppose that F is h-continuous. Thus using the notation of Theorem 67, the set T_∞ has zero μ^h-measure, by theorem 67, and so

$$F(T_\infty) = 0.$$

Now $I_0 \backslash T_\infty$ is the set of points x where

$$\bar{D}_h F(x) = \lim_{\delta \to 0} \sup_{\substack{x \in I \\ d(I) < \delta}} \frac{F(I)}{h(I)}$$

is finite. But this set is the union of the sets $L_{k,(1/n)}$, $n, k = 1, 2, \ldots$, where, as in theorem 65, $L_{k,(1/n)}$ is the set of points x with

$$\sup_{\substack{x \in I \\ d(I) \leqslant 1/n}} \frac{F(I)}{h(I)} \leqslant k.$$

By theorem 65 we know that the functions

$$G_{k,(1/n)}(t) = F(L_{k,(1/n)} \cap [0, t])$$

belong to $\mathscr{P}$ and are uniformly *Lip-h*; further the sets $L_{k,(1/n)}$ are Borel sets.

Let $L_1, L_2, \ldots$ be an enumeration of the sets $L_{k,(1/n)}$, $n, k = 1, 2, \ldots$, and write

$$D_i = L_i \backslash \{\bigcup_{j<i} L_j\}.$$

Then $T_\infty, D_1, D_2, \ldots$, is a sequence of disjoint Borel sets with union I_0. So, for all t with $0 \leqslant t \leqslant 1$,

$$\begin{aligned} F(t) &= F([0, t]) \\ &= F\left(\left\{T_\infty \cup \bigcup_{i=1}^{\infty} D_i\right\} \cap [(0, t]\right) \\ &= F(T_\infty \cap [0, t]) + \sum_{i=1}^{\infty} F(D_i \cap [0, t]) \\ &= \sum_{i=1}^{\infty} G_i(t), \end{aligned}$$

on writing

$$G_i(t) = F(D_i \cap [0, t]).$$

Now, for any a, b with $0 \leqslant a < b \leqslant 1$, we have

$$\begin{aligned} G_i(b) - G_i(a) &= F(D_i \cap [a, b]) \\ &\leqslant F(L_i \cap [a, b]) \\ &= F(L_{k,(1/n)} \cap [a, b]) \\ &= G_{k,(1/n)}(b) - G_{k,(1/n)}(a), \end{aligned}$$

for some choice of the integers n, k. As the functions $G_{k,(1/n)}$ is uniformly *Lip-h*, it follows that G_i is also uniformly *Lip-h*.

(*d*) Suppose that F is strongly h-continuous. Then using the notation of theorem 67 and the results of that theorem the set $T_+ \cup T_\infty$ has σ-finite μ^h-measure, and so

$$F(T_+ \cup T_\infty) = 0.$$

Now $T_0 = I_0 \backslash \{T_+ \cup T_\infty\}$ is the set of points x where

$$\bar{D}_h F(x) = \lim_{\delta \to 0} \sup_{\substack{x \in I \\ d(I) < \delta}} \frac{F(I)}{h(I)}$$

is zero. So we have $$T_0 = \bigcap_{k=1}^{\infty} \bigcup_{n=1}^{\infty} L_{(1/k),(1/n)},$$

where, as in theorem 65, $L_{(1/k)\,(1/n)}$ is the set of points x with

$$\sup_{\substack{x \in I \\ d(I) \leqslant 1/n}} \frac{F(I)}{h(I)} \leqslant \frac{1}{k}.$$

Clearly the sequence of sets

$$\bigcup_{n=1}^{\infty} L_{(1/k),(1/n)} \quad (k = 1, 2, \ldots),$$

is non-increasing.

Now the sets $L_{(1/k),(1/n)}$ are Borel subsets of I_0 and so are F-measurable. Further T_0 has finite F-measure. So, by the corollary to theorem 22 of §4.6 of Chapter 2, we can choose a strictly increasing sequence $n(1), n(2), \ldots,$ of positive integers, so that the sets

$$M_K = \bigcap_{k \geqslant K} \bigcup_{n \leqslant n(k)} L_{(1/k),(1/n)}$$

satisfy the conditions

$$F(M_K) \geqslant F(T_0) - 2^{-K+1},$$

for $K = 1, 2, 3, \ldots$.

Note that, as the set $L_{(1/k),(1/n)}$ does not decrease as n increases, we have

$$M_K = \bigcap_{k \geqslant K} L_{(1/k),(1/n(k))}.$$

Write $$D_i = M_i \backslash \{\bigcup_{j<i} M_j\}.$$

Then $D_1, D_2, \ldots$ is a sequence of disjoint Borel sets and

$$F(T_0) = F(I_0) \geqslant F\left(\bigcup_{i=1}^{\infty} D_i\right) \geqslant F(M_K) \geqslant F(T_0) - 2^{-K+1},$$

for all K. So
$$F\left(\bigcup_{i=1}^{\infty} D_i\right) = F(T_0) = F(I_0),$$
and
$$F\left(I_0 \backslash \bigcup_{i=1}^{\infty} D_i\right) = 0.$$
So, for all t with $0 \leqslant t \leqslant 1$,
$$\begin{aligned} F(t) &= F([0,t]) \\ &= F\left(\left\{I_0 \backslash \bigcup_{i=1}^{\infty} D_i\right\} \cap [0,t]\right) + \sum_{i=1}^{\infty} F(D_i \cap [0,t]) \\ &= \sum_{i=1}^{\infty} G_i(t), \end{aligned}$$
on writing,
$$G_i(t) = F(D_i \cap [0,t]).$$

Now consider any fixed positive integer i. If j is an integer with $j \geqslant i$ and a, b satisfy $0 \leqslant a < b \leqslant 1, b-a \leqslant 1/n(j)$, we have
$$\begin{aligned} G_i(b) - G_i(a) &= F(D_i \cap [A,b]) \\ &\leqslant F(M_i \cap [a,b]) \\ &\leqslant F(L_{(1/j),(1/n(j))} \cap [a,b]). \end{aligned}$$
By the assumption $d([a,b]) \leqslant 1/n(j)$, and the definition of the set $L_{(1/j),(1/n(j))}$, either
$$L_{(1/j),(1/n(j))} \cap [a,b] = \varnothing$$
so that $G_i(b) - G_i(a) = 0$, or this set is non-empty and
$$G_i(b) - G_i(a) \leqslant F([a,b]) \leqslant (1/j)\,h([a,b]).$$
In either case
$$G_i(b) - G_i(a) \leqslant (1/j)\,h(b-a). \tag{30}$$
Thus G_i is uniformly *lip-h*.

Corollary 1. *When F is strongly h-continuous, there is a function g of $\mathscr{H}_0$ with $h \prec g$ such that F has a convergent representation*
$$F(t) = \sum_{i=1}^{\infty} G_i(t) \quad (0 \leqslant t \leqslant 1),$$
the functions $G_i, i = 1, 2, \ldots$, belonging to $\mathscr{P}$ and being uniformly lip-g.

Proof. We use the notation of part (d) of the proof of the theorem. As the sequence $n(1), n(2), \ldots$ is a strictly increasing sequence of positive integers, we can choose a monotonic increasing continuous function λ with
$$\lim_{t \to 0} \lambda(t) = 0,$$

but $$\lambda(t) > 1/\sqrt{j} \quad (\text{for } t \geqslant 1/n(j+1)),$$

$j = 1, 2, \ldots$. Now, if i is a given positive integer, and a, b satisfy

$$0 \leqslant a < b \leqslant 1 \quad (b-a < 1/n(i)),$$

we can choose a positive integer j with $j \geqslant i$ and

$$1/n(j+1) < b-a \leqslant 1/n(j),$$

and use (30). Then

$$\begin{aligned} G_i(b) - G_i(a) &= (1/j)\, h(b-a) \\ &= (1/\sqrt{j})\,.\,(1/\sqrt{j})\, h(b-a) \\ &\leqslant (1/\sqrt{j})\, \lambda(1/n(j+1))\, h(b-a) \\ &\leqslant (1/\sqrt{j})\, \lambda(b-a)\, h(b-a) \\ &= (1/\sqrt{j})\, g(b-a), \end{aligned}$$

on writing $$g(t) = \lambda(t)\,.\,h(t).$$

Hence, g is a function of $\mathscr{H}_0$ with $h \prec g$, and each function G_i is uniformly *lip-g*.

***Corollary* 2.** *When F is strongly h-continuous, there is a function g of $\mathscr{H}_0$ with $h \prec g$ such that F is strongly g-continuous.*

Proof. The result follows immediately from corollary 1 and the theorem.

3.6. The Radon–Nikodym theorem depends on the concept of absolute continuity and of integration with respect to a measure. The first of these we will explain; but the second must be taken for granted, if this is to be the final section of this book.

***Definition* 39.** *If ϕ and ν are measures on a space Ω. the measure ϕ is said to be absolutely continuous with respect to the measure ν, if*: (a) *$\phi(\Omega)$ is finite*; (b) *all ν-measurable sets are ϕ-measurable*; *and* (c) *all sets E with $\nu(E) = 0$ have $\phi(E) = 0$.*

The Radon–Nikodym theorem (see, for example, Radon (1913), Munroe (1953), p. 196, or Saks (1937), p. 36) asserts that: *if ϕ and ν are measures on a space Ω of σ-finite ν-measure and ϕ is absolutely continuous with respect to ν, then there is a point function f defined on Ω, such that*

$$\phi(E) = \int_E f\,d\nu,$$

for all ν-measurable sets E.

This will enable us to prove the following theorem, already stated at the beginning of §3.

Theorem 69. *Suppose that $h \in \mathscr{H}_0$, that $F \in \mathscr{P}$, and that F is h-continuous. Then there is a point function f defined on I_0, a function J of $\mathscr{P}$ and a function g of $\mathscr{H}_0$, such that*

$$F(E) = \int_E f d\mu^h + J(E),$$

for all Borel sets E, where J is strongly g-continuous and $h \prec g$.

Proof. We use the notation of theorem 67. As $\mu^h(T_\infty) = 0$, by theorem 67, and F is h-continuous we have $F(T_\infty) = 0$. As T_0, T_+ and T_∞ are F-measurable, we have

$$F(E) = F(E \cap T_+) + F(E \cap T_0)$$

for all sets E.

Define the set functions ϕ, ν and the function J of $\mathscr{P}$ by the formulae

$$\phi(E) = F(E \cap T_+), \tag{31}$$

$$\nu(E) = \mu^h(E \cap T_+), \tag{32}$$

$$J(t) = F(T_0 \cap [0, t]) \quad (0 \leqslant t \leqslant 1). \tag{33}$$

Then ν is clearly a measure on I_0, and, by (*b*) of theorem 67, I_0 is of σ-finite ν-measure. Further all Borel sets are ν-measurable. If E is any set, we can choose a Borel set B with

$$E \cap T_+ \subset B, \qquad \mu^h(E \cap T_+) = \mu^h(B).$$

Then $C = B \cup \{I_0 \backslash T_+\}$ is a Borel set with

$$E \subset C, \qquad \nu(E) = \mu^h(E \cap T_+) = \nu(C).$$

So any ν-measurable set E of finite ν-measure takes the form

$$E = C \backslash \{C \backslash E\},$$

where C is Borel and $C \backslash E$ has zero ν-measure. But, if a set D has zero ν-measure, $\mu^h(D \cap T_+) = 0$, so that

$$\phi(D) = F(D \cap T_+) = 0, \tag{34}$$

by (*c*) of theorem 67. Hence C being Borel and $C \backslash E$ being of zero ϕ-measure will both be ϕ-measurable. Thus each ν-measurable set is ϕ-measurable, and, by (34), each set D of zero ν-measure has zero ϕ-measure. So ϕ is absolutely continuous with respect to ν.

Applying the Radon–Nikodym theorem to ϕ and ν we obtain a point function f_0 on I_0 such that, for all ν-measurable sets E,

$$\phi(E) = \int_E f_0 d\nu.$$

Write
$$f(t) = f_0(t) \quad \text{if} \quad t \in T_+,$$
$$f(t) = 0 \quad \text{if} \quad t \notin T_+.$$

Then it follows, from (31), (32) and the theory of integration, that

$$F(E \cap T_+) = \phi(E) = \int_E f_0 d\nu = \int_E f d\mu^h.$$

We turn to the function J of $\mathscr{P}$ defined by (33). By (d) of theorem 67, if E has σ-finite μ^h-measure,

$$J(E) = F(E \cap T_0) = 0.$$

So J is strongly h-continuous. By corollary 2 to theorem 68, there is a function g of $\mathscr{H}_0$ with $h \prec g$ such that J is strongly g-continuous.

Since, for all Borel sets E,

$$F(E) = F(E \cap T_+) + F(E \cap T_0) = \int_E f d\mu^h + J(E),$$

this completes the proof.

BIBLIOGRAPHY

BEARDON, A. F.: 1965*a*, On the Hausdorff dimension of general Cantor sets, *Proc. Camb. phil. Soc.* **61**, 679–94; 1965*b*, The Hausdorff dimension of singular sets of properly discontinuous groups in N-dimensional space, *Bull. Am. math. Soc.* **71**, 160–15; 1966, The Hausdorff dimension of singular sets of properly discontinuous groups, *Am. J. Math.* **88**, 722–36.

BESICOVITCH, A. S.: 1928, On the fundamental geometric properties of linearly measurable plane sets, *Math. Annln*, **98**, 422–64; 1929, On linear sets of fractional dimensions, *Math. Annln*, **101**, 161–93; 1934*a*, Sets of fractional dimensions II: On the sum of digits of real numbers represented in the dyadic system, *Math. Annln*, **110**, 321–9; 1934*b*, Sets of fractional dimensions III: Sets of points of non-differentiability of absolutely continuous functions and of divergence of Fejer sums, *Math. Annln*, **110**, 331–5; 1934*c*, Sets of fractional dimensions IV: On rational approximation to real numbers, *J. Lond. math. Soc.* **9**, 126–31; 1934*d*, Concentrated and rarified sets of points, *Acta Math., Stock.* **62**, 289–300; 1938, On the fundamental geometric properties of linearly measurable plane sets of points II, *Math. Annln.* **115**, 296–329; 1939, On the fundamental geometric properties of linearly measurable plane sets of points III, *Math. Annln*, **116**, 349–57; 1942, A theorem on s-dimensional measure of sets of points, *Proc. Camb. phil. Soc.* **38**, 24–7; 1948*a*, On distance sets, *J. Lond. math. Soc.* **23**, 9–14; 1948*b*, On surfaces of minimum area, *Proc. Camb. phil. Soc.* **44**, 313–34; 1949*a*, Parametric surfaces I: Compactness, *Proc. Camb. phil. Soc.* **45**, 5–13; 1949*b*, Parametric surfaces II: Lower semi-continuity of the area, *Proc. Camb. phil. Soc.* **45**, 14–23; 1948*c*, Parametric surfaces III: On surfaces of minimum area, *J. Lond. math. Soc.* **23**, 241–6; 1949*c*, Parametric surfaces IV: The integral formula for the area, *Q. J. Math., Oxford*, **23**, 241–6; 1950, Parametric surfaces, *Bull. Am. math. Soc.* **56**, 288–96; 1952, On existence of subsets of finite measure of sets of infinite measure, *Indagat. math.* 14, 339–44; 1954*a*, Parametric surfaces. III I: Surfaces of minimal area, *Indagat. math.* **16**, 169–75; 1954*b*, On approximations in measure to Borel sets by F_σ-sets, *J. Lond. math. Soc.* **29**, 382–3; 1956*a*, On density of perfect sets, *J. Lond. math. Soc.* **31**, 48–53; 1956*b*, On the definition of tangents to sets of infinite linear measure, *Proc. Camb. phil. Soc.* **52**, 20–9; 1957*a*, On density of linear sets, *J. Lond. math. Soc.* **32**, 170–8; 1957*b*, Analysis of tangential properties of curves of infinite length, *Proc. Camb. phil. Soc.* **53**, 69–72; 1960, Tangential properties of sets and arcs of infinite linear measure, *Bull. Am. math. Soc.* **66**, 353–9; 1963*a*, On the set of directions of linear segments on a convex surface, *Proc. Symp. Pure Math.*, vol. VII, *Convexity*, Am. math. Soc., Providence, R.I., 1963, 24–5; 1963*b*, On singular points of convex surfaces, *Convexity*, 21–3; 1963*c*, A problem on measure, *Proc. Camb. phil. Soc.* **59**, 251–3; 1964*a*, On one-sided densities of arcs of positive two-dimensional measure, *Proc. Camb. phil. Soc.* **60**, 517–24; 1964*b*, On fundamental geometric properties of plane line-sets, *J. Lond. math. Soc.* **39**, 441–8; See also *Math. Rev.* **30**, vol. II, p. 414; 1967, On linear sets of points of fractional dimension II, *J. Lond. math. Soc.* **43**, 548–50; *Geometry of sets of points*, to appear.

Besicovitch, A. S. and D. S. Miller: 1948, On the set of distances between the points of a Carathéodory linearly measurable plane point set, *Proc. Lond. math. Soc.* (2), **50**, 305–16.

Besicovitch, A. S. and P. A. P. Moran: 1945, The measure of product and cylinder sets, *J. Lond. math. Soc.* **20**, 110–20.

Besicovitch, A. S. and I. J. Schoenberg: 1961, On Jordan arcs and Lipschitz classes of functions defined on them, *Acta Math.* **106**, 113–36.

Besicovitch, A. S. and S. J. Taylor: 1954, On the complementary intervals of a linear closed set of zero Lebesgue measure, *J. Lond. math. Soc.* **29**, 449–59; 1952, On the set of distances between points of a general metric space, *Proc. Camb. phil. Soc.* **48**, 209–14.

Besicovitch, A. S. and H. D. Ursell: 1937, Sets of fractional dimensions V: On dimensional numbers of some continuous curves, *J. Lond. math. Soc.* **12**, 18–25.

Besicovitch, A. S. and G. Walker: 1931, On the density of irregular linearly measurable sets of points, *Proc. Lond. math. Soc.* **32**, 142–53.

Best, E.: 1939, A closed dimensionless linear set, *Proc. Edinb. math. Soc.* (2), **6**, 105–8; 1940*a*, A theorem on Hausdorff measure, *Q. J. Math. Oxford*, **11**, 243–8; 1940*b*, On sets of fractional dimensions, *Proc. Camb. phil. Soc.* **36**, 152–9; 1941, On sets of fractional dimensions II, *Proc. Camb. phil. Soc.* **37**, 127–33; 1942, On sets of fractional dimensions III, *Proc. Lond. math. Soc.* (2), **47**, 436–54.

Beyer, W. A.: 1962, Hausdorff dimensions of level sets of some Rademacher series, *Pacif. J. Math.* **12**, 35–46.

Billingsley, P.: 1960, Hausdorff dimension in probability theory, *Illinois J. Math.* **4**, 187–209; 1961, Hausdorff dimension in probability theory II, *J. Math.* **5**, 291–8.

Blaschke, W. : 1916, *Kreis and Kugel*, Leipzig, p. 169.

Bledsoe, W. W. and A. P. Morse: 1955, Product measures, *Trans. Am. math. Soc.* **79**, 173–215; 1963, A topological measure construction, *Pacif. J. Math.* **13**, 1067–84.

Blumenthal, R. M. and R. K. Getoor: 1960*a*, Some theorems on stable processes, *Trans. Am. math. Soc.* **95**, 263–73; 1960*b*, A dimension theorem for sample functions of stable processes, *Illinois J. Math.* **4**, 370–5; 1961, Sample functions of stochastic processes with stationary independent increments, *J. Math. Mech.* **10**, 493–516; 1962, The dimension of the set of zeros and the graph of a symmetric stable process, *Illinois J. Math.* **6**, 308–16; 1964, Local time for Markov processes, *Z. Wahrscheinlichkeitstheorie und Verw. Gebiete*, **3**, 50–74.

Borel, E.: 1895, Sur quelques points de la théorie des fonctions, *Ann. École. Norm. sup* (3), **12**, 9–55; 1898, *Leçons sur la théorie des fonctions*, Gauthier-Villars, Paris, 136; 1909, Les probabilitiés dénombrables et leurs applications arithmétique, *Rc. Circ. mat. Palermo*, **27**, 247–70; 1940, *Selecta de M. Émile Borel*, Gauthier-Villars, Paris, p. 418.

Carathéodory, C.: 1914, Über das lineare Mass von Punktmengeneine Verallgemeinerung des Längenbegriffs *Nach. Ges. Wiss. Göttingen*, pp. 404–26; 1918, *Voresungen über reelle Funktionen*, Teubner, Leipzig–Berlin.

Carleson, L.: 1958, On the connection between Hausdorff measures and capacity, *Ark. Mat.* **3**, 403–6; 1967, *Selected problems of Exceptional Sets*, Van Nostrand, Princeton, p. 151.

CHOQUET, G.: 1947, Ensembles singuliers et structure des ensembles mesurable pur les mesures de Hausdorff, *Bull. Soc. math. France*, **74**, 1–14.

CIESIELSKI, Z. and S. J. TAYLOR.: 1962, First passage times and sojourn times for Brownian motion in space and the exact Hausdorff measure of the sample path, *Trans. Am. math. Soc.* **103**, 434–50.

CIGLER, J.: 1961, Hausdorffische Dimensionen spezieller Punktmengen, *Math. Z.* **76**, 22–30.

DAVENPORT, H: 1952, *The Higher Arithmetic*, Hutchinson, London, p. 172.

DAVIES, ROY O.: 1952, Subsets of finite measure in analytic sets, *Indagat. math.* **14**, 488–9; 1956*a*, A property of Hausdorff measure, *Proc. Camb. phil. Soc.* **52**, 30–2; 1956*b*, Non σ-finite closed subsets of analytic sets, *Proc. Camb. phil. Soc.* **52**, 174–7; 1965, A regular line-set covering finite measures, *J. Lond. math. Soc.* **40**, 503–8; 1968, A theorem on the existence of non-σ-finite subsets, *Mathematika*, **15**, 60–2; 1969, Measures of Hausdorff type, *J. Lond. math. Soc.* (2), **1**, 30–4; 1970, Increasing sequences of sets and Hausdorff measure, *Proc. Lond. math. Soc.* (3), **20**, 222–36; 197?. Some remarks on the Kakeya problem, *to appear*.

DAVIES, ROY O., J. M. MARSTRAND and S. J. TAYLOR: 1960, On the intersections of transforms of linear sets, *Colloq. Math.* **7**, 237–43.

DAVIES, ROY O. and C. A. ROGERS: 1969, The problem of subsets of finite positive measure, *Bull. Lond. math. Soc.* **1**, 47–54.

DICKINSON, D. R.: 1939, Study of extreme cases with respect to the densities of irregular linearly measurable plane set of points, *Math. Annln*, **116**, 359–73.

DVORETZKY, A.: 1948, A note on Hausdorff dimension functions, *Proc. Camb. phil. Soc.* **44**, 13–16.

DVORETZKY, A., P. ERDŐS and S. KAKUTANI: 1950, Double points of paths of Brownian motion in n-space, *Acta Sci. Math. Szeged*, **12**, 75–81; 1954, Multiple points of paths of Brownian motion, *Bull. Res. Coun. Israel*, **3**, 364–71.

DVORETZKY, A., P. ERDŐS, S. KAKUTANI and S. J. TAYLOR: 1957, Triple points of Brownian paths in 3-space, *Proc. Camb. phil. Soc.* **53**, 856–62.

DVORETZKY, A. and S. KAKUTANI: 1958, Points of multiplicity c of plane Brownian motion, *Bull. Res. Counc. Israel*, F, **7** F, 175–80.

EGGLESTON, H. G.: 1949*a*, Note on certain s-dimensional sets, *Fund. Math.* **36**, 40–3; 1949*b*, Homeomorphisms, of s-sets, *J. Lond. math. Soc.* **24**, 181–90; 1949*c*, The fractional dimension of a set defined by decimal properties *Q. J. Math. Oxford*, **20**, 31–6; 1950*a*, The Besicovitch dimension of Cartesian product sets, *Proc. Camb. phil. Soc.* **46**, 383–6; 1950*b*, A property of Hausdorff measure, *Duke Math. J.* **17**, 491–8; 1950*c*, A characteristic property of Hausdorff measure, *J. Lond. math. Soc.* **25**, 39–46; 1950*d*, A geometrical property of sets of fractional dimension, *Q. J. Math. Oxford* (2) **1**, 81–5; 1951, Correction to 'A property of Hausdorff measure', *Duke Math. J.* **18**, 593; 1952, Sets of fractional dimension which occur in some problems of number theory, *Proc. Lond. math. Soc.* (2), **54**, 42–93; 1953*a*, On closest packing by equilateral triangles, *Proc. Camb. phil. Soc.* **49**, 26–30; 1953*b*, A correction to a paper on the dimension of Cartesian product sets, *Proc. Camb. phil. Soc.* **49**, 437–40; 1954, A measureless one-dimensional set, *Proc. Camb. phil. Soc.* **50**, 391–3; 1954, Two measure properties of

Cartesian product sets, *Q. J. Math. Oxford*, (2), **5**, 108–15; 1958*a*, *Convexity*, Cambridge, p. 136; 1958*b*, On measureless sets, *Proc. Lond. math. Soc.* (3), **8**, 631–40; 1958*c*, Tangential properties of Fréchet surfaces, *Proc. Camb. phil. Soc.* **54**, 187–96; 1958*d*, On the projection of a plane set of finite linear measure, *Acta Math. Stock.* **99**, 53–91.

EGOROFF, D. T.: 1911, Sur les suites des fonctions mesurables, *C.r. hebd. Séanc. Acad. Sci., Paris*, **152**, 244–6.

ERDŐS, P.: 1940, The dimension of the rational points in Hilbert space, *Ann. Math.* (2), **41**, 734–6; 1946, On the Hausdorff dimension of some sets in Euclidean space, *Bull. Am. math. Soc.* **52**, 107–9.

ERDŐS, P., H. KESTELMAN, and C. A. ROGERS: 1963, An intersection property of sets with positive measure, *Colloquium math.* **11**, 75–80.

ERDŐS, P., C. A. ROGERS and S. J. TAYLOR: 1960, Scales of functions, *J. Austral. math. Soc.* **1**, 396–418.

ERDŐS, P. and S. J. TAYLOR,: 1960*a*, Some problems concerning the structure of random walk paths, *Acta math. Hung.* **11**, 137–62; 1960*b*, Some intersection properties of random walk paths *Acta math. Hung.* **11**, 231–48; 1961, On the Hausdorff measure of Brownian paths in the plane, *Proc. Camb. phil. Soc.* **57**, 209–22; 1963; The Hausdorff measure of the intersection of sets of positive Lebesgue measure, *Mathematika*, **10**, 1–9.

EROHIN, V.: 1958, The connection between metric dimension and harmonic capacity (Russian), *Usp. mat. Nauk*, **13**, no. 6, 81–8.

ESTERMANN, T.: 1926, Über Carathéodory's und Minkowski's Verallgemeinerung des Längenbegriffs, *Abh. math. Semin. Univ. Hamburg*, **4**, 73–116.

EWALD, G., D. G. LARMAN and C. A. ROGERS: 1970, The directions of the line segments and of the *r*-dimensional balls on the boundary of a convex body in Euclidean space, *Mathematika*, **17**, 1–20.

FEDERER, H. : 1951, Hausdorff measure and Lebesgue area, *Proc. Nat. Acad. Sci. U.S.A.* **37**, 90–4; 1952, Measure and area, *Bull. Am. math. Soc.* **58**, 306–78; 1969, *Geometric measure theory, Die Grundlehren der math. Wiss.* p. 153, Springer, Berlin, 676.

FREILICH, G.: 1950, On the measure of cartesian product sets, *Trans. Am. math. Soc.* **69**, 232–75; 1965; Carathéodory measure of cylinders, *Trans. Am. math. Soc.* **114**, 384–400.

GIERL, A.: 1959, Über das Hausdorffsche Mass gewisser Punktmengen in der Zifferntheorie, *J. reine angew. Math.* **202**, 183–95.

GILLIS, J.: 1934, On the projection of irregular linearly measurable plane sets of points, *Proc. Camb. phil. Soc.* **30**, 47–54; 1935, A theorem on irregular linear measurable sets of points, *J. Lond. math. Soc.* **10**, 234–40.

GLIVENKO, E. V.: 1956, On measures of the Hausdorff type (Russian), *Mat. Sb. N.S.* **39**, (81), 423–32.

GOOD, I. J.: 1941, The fractional dimensional theory of continued fractions, *Proc. Camb. phil. Soc.* **37**, 199–228.

GROSS, W.: 1918*a*, Über das Flächenmass von Punktmengen, *Monat. Math. Physik*, **29**, 145–176; 1918*b*, Über das lineare Mass von Punktmengen, *Monat. Math. Physik*, **29**, 177–93.

HALMOS, P. R.: 1950, *Measure theory*, Van Nostrand, Princeton, 304.

HARDY, G. H. and E. M. WRIGHT: 1954, *Theory of Numbers*, 3rd edn., Oxford University Press, Oxford, p. 419.

HAUSDORFF, F.: 1919, Dimension und äusseres Mass, *Math. Ann.* **79**, 157–79.

HAWKES, J.: 1970, Measure function properties of the asymmetric Cauchy process, *Mathematika,* **17**, 68–78.

HIRST, K. E.: 1967, The Apollonian packing of circles, *J. Lond. math. Soc.* **42**, 281–91; 1970, Fractional dimension theory of continued fractions, *Q. J. Math., Oxford,* (2), **21**, 29–35.

HUREWICZ, W. and H. WALLMAN: 1941, *Dimension Theory,* Princeton University Press, Princeton, p. 165.

JARNÍK, V.: 1928, Zur metrischen Theorie der diophantischen Approximationen, *Pr. mat-fiz.* **36**, 91–106.

KAMETANI, S.: 1941, Theorems on interval functions and h-measures, *Jap. J. Math.* **17**, 533–9; 1942, On some properties of Hausdorff's measure and the concept of capacity in generalized potentials, *Proc. Imp. Acad. Tokyo,* **18**, 617–25, and **20**, 15; 1945, On Hausdorff's measures and generalized capacities with some of their applications to the theory of functions, *Jap. J. Math.* **19**, 217–57.

KAUFMAN, R.: 1968, Dimensions of projections, *Mathematika,* **15**, 153–5; 1969, An exceptional set for Hausdorff dimension, *Mathematika,* **16**, 57–8.

KINNEY, J. R.: 1958, Singular functions associated with Markov chains, *Proc. Am. math. Soc.* **9**, 603–8; 1960, Note on a singular function of Minkowski, *Proc. Am. math. Soc.* **11**, 788–94.

KINNEY, J. R. and T. S. PITCHER: 1964, The dimension of the support of a random distribution function, *Bull. Am. math. Soc.* **70**, 161–4; 1965/6, The dimension of some sets defined in terms of f-expansions, *Z. Wahrschienlichkeitsheorie und Verw. Gebiete,* **4**, 293–315; 1966*a*, The Hausdorff–Besicovitch dimension of the level sets of Perron's modular function, *Trans. Am. math. Soc.* **124**, 122–30; 1966*b*, Dimensional properties of a random distribution function on the square, *Ann. math. Statist.* **37**,849–54; 1969, Perron's modular function, *Can. J. Math.* **21**, 808–16,

KLINE, S. A.: 1945, On curves of fractional dimensions, *J. Lond. math. Soc.* **20**, 79–86.

KNOWLES, J. D.: 1966, On the construction of measures, *Mathematika,* **13**, 60–8; 1967, On the existence of non-atomic measures, *Mathematika,* **14**, 62–7.

KURATOWSKI, K.: 1966, *Topology* (Academic Press, New York).

LARMAN, D. G.: 1965, The approximation of G_δ-sets, in measure, by F_σ-sets, *Proc. Camb. phil. Soc.* **61**, 105–7; 1966*a*, A note on the Besicovitch dimension of the closest packing of spheres in R_n, *Proc. Camb. phil. Soc.* **62**, 193–5; 1966*b*, On the exponent of convergence of a packing of spheres *Mathematika,* **13**, 57–9; 1966*c*, Subsets of given Hausdorff measure in connected spaces, *Q. J. Math. Oxford* (2), **17**, 239–43; 1967*a*, On the Besicovitch dimension of the residual set of arbitrarily packed disks in the plane, *J. Lond. math. Soc.* **42**, 292–302; 1967*b*, A new theory of dimension, *Proc. Lond. math. Soc.* (3), **17**, 178–92; 1967*c*, On Hausdorff measure in finite-dimensional compact metric spaces, *Proc. Lond. math. Soc.* (3), **17**, 193–206; 1967*d*, On selection of non-σ-finite subsets, *Mathematika,* **14**, 161–4 1967*e*, On the convex measure of product and cylinder sets, *J. Lond. math. Soc.* **42**, 447–55; 1968, On the right lower circular density of two-dimensional Jordan curves in the plane, *Proc. Camb. phil. Soc.* **64**, 67–70.

LARMAN, D. G. and D. J. WARD: 1966, On convex sets and measures, *Proc. Camb. phil. Soc.* **62**, 33–41.

LEBESGUE, H.: 1904, *Leçons sur l'intégration et la recherche des fonctions primitives*, Gauthier–Villars, Paris, p. 342.

LÉVY, P.: 1948, *Processus stochastiques et mouvement brownien*, Gauthier–Villars, Paris, p. 365; 1951, La mesure de Hausdorff de la courbe du mouvement brownien à n dimensions, *C.r. hebd. Séanc. Acad. Sci., Paris*, **233**, 600–2; 1954; *Le mouvement brownien, Mém. Sci. Math.* no. **126**, Gauthier–Villars, Paris, p. 84.

LUSIN, N.: 1917, Sur la classification de M. Baire, *Comptes Rendue*, **164**, 91–4.

MCMINN, T. J.: 1960, On the line segments of a convex surface in E_3, *Pacif. J. Math.* **10**, 943–6.

MARSTRAND, J. M.: 1954*a*, The dimension of the Cartesian product sets, *Proc. Camb. phil. Soc.* **50**, 198–202; 1954*b*, Some fundamental geometrical properties of plane sets of fractional dimensions, *Proc. Lond. math. Soc.* **4**, 257–302; 1954*c*, Circular density of plane sets, *J. Lond. math. Soc.* **30**, 238–46; 1961, Hausdorff two-dimensional measure in 3-space, *Proc. Lond. math. Soc.* (3), 11, 91–108; 1964, The (ϕ, s)-regular subsets of n space, *Trans. Am. math. Soc.* **113**, 369–92.

MELZAK, Z. A.: 1966, On infinite packings of disks, *Can. J. Math.* **18**, 838–52.

MICKLE, E. J.: 1955, Lebesgue area and Hausdorff measure, *Rc. Circ. mat. Palermo* (2), **4**, 205–18; 1958, On a closure property of measurable sets, *Proc. Am. math. Soc.* **9**, 688–9.

MICKLE, E. J. and T. RADO: 1958, Density theorems for outer measures in n-space, *Proc. Am. math. Soc.* **9**, 433–9.

MORAN, P. A. P.: 1946, Additive functions of intervals and Hausdorff measure, *Proc. Camb. phil. Soc.* **42**, 15–23; 1949; On plane sets of fractional dimensions, *Proc. Lond. math. Soc.* (2), **51**, 415–23; 1954, The translations of linear sets of fractional dimensions, *Proc. Camb. phil. Soc.* **50**, 634–6.

MORGAN, G. W.: 1935, The density directions of irregular measurable sets, *Proc. Lond. math. Soc.* **38**, 481–94.

MUNROE, M. E.: 1953, *Introduction to Measure and Integration*, Addison–Wesley, Reading, Mass., U.S.A. p. 310.

NEMITZ, W. C.: 1961, On a decomposition theorem for measures in Euclidean n-space, *Trans. Am. math. Soc.* **98**, 306–33.

RADON, J.: 1913, Theorie und Anwendungen der absolut additiven Mengenfunktionen, *Sber. Akad. Wiss. Wien*, **122**, 1295–438.

RAVETZ, J.: 1954, The Denjoy theorem and sets of fractional dimension, *J. Lond. math. Soc.* **29**, 88–96.

RAY, D.: 1963, Sojourn times and exact Hausdorff measure of the sample path for planar Brownian motion, *Trans. Am. math. Soc.* **106**, 436–44.

REIFENBERG, E. R.: 1951, Parametric surfaces I: Area, *Proc. Camb. phil. Soc.* **47**, 687–98; 1952*a*, Parametric surfaces II: Tangential properties, *Proc. Camb. phil. Soc.* **48**, 46–69; 1952*b*, Parametric surfaces III: The problem of Geocze, *Q. J. Math. Oxford* (2), **3**, 227–34; 1952*c* Parametric surfaces IV: The generalized Plateau problem, *J. Lond. math. Soc.* **27**, 448–56; 1955, Parametric surfaces V: Area II, *Proc. Lond. math. Soc.* (3), **5**, 342–57; 1960*a*, Solution of the Plateau problem for m-dimensional surfaces of varying topological type, *Bull. Am. math. Soc.* **66**, 312–13; 1960*b*, Solution of the Plateau problem for m-dimensional surfaces of varying topological type, *Acta Math., Stock.* **104**, 1–92; 1962, On the tangential properties of surfaces, *Bull. Am. math. Soc.* **68**, 213–6; 1964, An epiperimetric inequality related to the analyticity of minimal surfaces: On the analyticity of minimal surfaces, *Ann. Math.* **80**, 1–21.

ROGERS, C. A.: 1961, Dense scales of functions, *J. Austral. math. Soc.* **2**, 137–42; 1961, Uniform set functions, *Proc. Royal Soc. Lond.* A, **263**, 149–60; 1962, Sets non-σ-finite for Hausdorff measures, *Mathematika*, **9**, 95–103; 1964, Some sets of continued fractions, *Proc. Lond. math. Soc.* (3), **14**, 29–44.

ROGERS, C. A. and M. SION: 1963, On Hausdorff measures in topological spaces, *Monat. Math.* **67**, 269–78.

ROGERS, C. A. and S. J. TAYLOR: 1959, Additive set functions in Euclidean space, *Acta Math., Stock.* **101**, 273–302 [but §6 is incorrect, see part II below]; 1961, Functions continuous and singular with respect to a Hausdorff measure, *Mathematika*, **8**, 1–31; 1962, On the law of the iterated logarithm, *J. Lond. math. Soc.* **37**, 145–51; 1963, Additive set functions in Euclidean space II, *Acta Math., Stock.* **109**, 207–40.

SAKS, S.: 1937, *Theory of the Integral*, 2nd edn., Hafner, New York, p. 347.

SALEM, R.: 1951, On singular monotonic functions whose spectrum has a given Hausdorff dimension, *Ark. Math.* **1**, 353–65.

SHERMAN, S.: 1942, A comparison of linear measures in the plane, *Duke Math. J.* **9**, 1–9.

SIERPIŃSKI, W.: 1927, Sur la densité linéaire des ensembles plans, *Fund. Math.* **9**, 172–85; 1956, *General Topology*, Toronto, p. 290.

SION, M. and D. SJERVE: 1962, Approximation properties of measures generated by continuous set functions, *Mathematika*, **9**, 145–56.

SION, M. and R. C. WILLMOTT.: 1966, Hausdorff measures on abstract spaces, *Trans. Am. math. Soc.* **123**, 275–309.

SMORODINSKY, M.: 1968, The capacity of a general noiseless channel and its connection with Hausdorff dimension, *Proc. Am. math. Soc.* **19**, 1247–54.

STEINHAUS, H.: 1920, Sur les distances des points des ensembles de mesure positive, *Fund. Math.* **1**, 93–104.

TAYLOR, S. J.: 1952, On Cartesian product sets, *J. Lond. math. Soc.* **27**, 295–304; 1953, The Hausdorff α-dimensional measure of Brownian paths in n-space, *Proc. Camb. phil. Soc.* **48**, 31–9; 1955*a*, On the Hausdorff measure of linear $\mathscr{A}$-sets, *Nieuw Arch. Wisk.* (3), **3**, 6–12; 1955*b*, The α-dimensional measure of the graph and set of zeros of a Brownian path, *Proc. Camb. phil. Soc.* **51**, 265–74; 1961, On the connexion between Hausdorff measures and generalized capacity, *Proc. Camb. phil. Soc.* **57**, 524–31; 1964, The exact Hausdorff measure of the sample path for planar Brownian motion, *Proc. Camb. phil. Soc.* **60**, 253–8; 1966, Multiple points for the sample paths of the symmetric stable processes, *Z. Wahrscheinlichkeitstheorie und Verw. Gebiete*, **5**, 247–64; 1967, Sample path properties of a transient stable process, *J. Math. Mech.* **16**, 1229–46.

TAYLOR, S. J. and J. G. WENDEL: 1966, The exact Hausdorff measure of the zero set of a stable process, *Z. Wahrscheinlichkeitstheorie und Verw. Gebiete*, **6**, 170–80.

VITALI, G.: 1908, Sui gruppi di punti e sulle funzioni di variabili reali, *Atti Accad. Sci., Torino*, **43**, 75–92.

WALKER, G.: 1929, On the density of irregular linearly measurable plane sets, *Proc. Lond. math. Soc.* (2), **30**, 481–99.

WARD, D. J.: 1964, A counterexample in area theory, *Proc. Camb. phil. Soc.* **60**, 821–45; 1967, The measure of cylinder sets, *J. Lond. math. Soc.*

42, 401–8; 1970*a*, A set of plane measure zero containing all finite polygonal arcs, *Can. J. Math.* to appear; 1970*b*, Some dimensional properties of generalized difference sets, *Mathematika*, **17**, to appear.

WEGMANN, H.: 1969*a*, Die Hausdorff-Dimension von kartesischen Produktmengen in metrischen Räumen, *J. reine angew. Math.* **234**, 163–71; Die Hausdorff-Dimension von Mengen reeller Zahlenfolgen, *J. reine angew. Math.* **235**, 20–8.

WILLMOTT, R. C.: 1967; Some properties of Hausdorff measures on uniform spaces, *Proc. Lond. math. Soc.* **17**, 513–29.

INDEX

(*Numbers in italics indicate pages where the concepts are introduced, or where theorems are stated*)